Robots and Telerobots in Space Applications

© 1991, 2011, 2017

Patrick H. Stakem

2nd edition
Number 3 in the Robot Series

Table of Contents

Introduction

This book will help you understand the fundamentals of robotics and telerobotics for the space environment. It will point out what the robotic systems can and can't do. Examples of systems, case study's, and design examples will be presented.

We will review the basics and definitions of robotic and telerobotic systems, as well as the unique characteristics of the space environment to determine where the trade-offs lie. We will compare and contrast with underwater, military, commercial, hazmat and other terrestrial systems. We will not discuss CAD/CAM or manufacturing, which probably makes up 90% of the applications of robotics on this planet.

We will review system level components, and discuss sensors, power sources, actuators, and computation and communication systems. Actuators will include tools and grippers. Necessarily, we will discuss simulation, task planning, simulation, guided autonomy, and autonomous systems, as well as system models.

"Man is the lowest cost, 150-pound, non-linear, all-purpose computer system that can be mass-produced by unskilled labor." Attributed to a NASA official.

Today's robot systems are deaf, nearly blind and stupid. Yet, we expect them to operate safely in an unstructured environment. But, they are getting better, as technology advances.

Robots are handicapped in terms of mobility and manipulation, sensory input, cognitive processing, learning, and the application of experience. However, they have better computational capability, better communications capability, fewer environmental constraints, and, perhaps, fewer ethical issues. (Leaving aside the issue of military armed robots).

What are the problem areas in the applications of robots? Accuracy, which has both a mechanical and a sensor component; dynamic performance, a speed/dexterity trade-off, addressed by more robust control algorithms, sensor systems in terms of integration and processing; interactive control, starting with modeling, standardization and modularization. None of these are insurmountable problems, and the advance of technology addresses all.

Robotic systems need better world models. They need to integrate and fuse sensor data into a better view of the world around them. They need more reasonableness assumptions, and a-priori knowledge of the physical world. They need flexibility of response. Learning from experience would be a major asset. What they need, then, is better software.

This edition has been extensively updated and revised. New material has been added, and the Mars section has been expanded to include the latest results. Not all the robot projects in space are covered, but a significant cross-section of them are.

In one sense, all spacecraft are robots, but here we restrict the discussion to units that can move around, explore, and fix things.

The Author

The author spent 42 years in support of numerous NASA spaceflight mission, on this planet and others. He teaches for the Johns Hopkins University, Whiting School of Engineering, Engineering for Professionals Program. He has taught for Capitol Technology University, Engineering and Computer Science Department, and Loyola University in Maryland, Graduate Department of Computer Science.

He began his aerospace career at Fairchild Industries, and overlapped with Dr. Wernher von Braun, so, technically, he was a member of the von Braun rocket team. He has received the NASA Shuttle Program Managers Commendation Award, two NASA Group Achievement Awards, Certificate of Appreciation from the NASA Earth Science Technology Office, and the NASA Apollo-Soyuz Test Program Award. He did extensive work on the Flight Telerobotic Servicer Mission. He served on AIAA's Committee on Standards for Automation and Robotics for TransOrbital, Lunar and Mars Base. He worked on the Solar Maximum Repair Mission, and briefly on the Hubble Space Telescope repair missions.

Definitions – what is a robot, what is a telerobot?

At the very top level, we need a strategic plan. From the objectives of this plan, a sequence of sub-plans and operations can be derived. We also need contingency cases and the ability to re-plan, when reality diverges from expectations. The robot needs to be able to, at some level, plan and execute, then evaluate according to success criteria. Task decomposition into relevant sub-tasks is an area that is essential for the more advanced telerobotic systems.

Telerobots are designed to operate in distant or remote areas. They extend the operating envelope of human workers in space and/or time. For example, an underwater telerobot does not need to surface to replenish its air supply, and when it does resurface, it doesn't have to do it slowly to avoid "the bends." A robot on the International Space Station can be kept outside, not needing to "suit-up" before attending to repair scenarios. Telerobots extend the domain where useful or critical work can be done. They reduce human labor requirements, and reduce human exposure to hazardous environments. Space is certainly a hazardous environment for humans. Besides space, telerobots are currently used on land, sea, in the air, and under the sea.

Telepresense is the extension of a human's senses to a remote location. This includes enhancement of the senses, such as a radiation detector, or night vision. Various aspects of the workstation design for the humans determine the effectiveness of the overall system. For example, should the cameras on the robotic system be slaved to the head motion of the operator? Our tools tend to have our capabilities in mind, so we work best with systems like us, bilaterally symmetric, for example. Having a three-armed robotic device, even though it would be very useful in many tasks, would be difficult for the human operate to get used to. From the task standpoint, a big, strong left arm and two dexterous right arms might be the right choice, Telecontrol refers to the manipulation of the remote system by the human operator, either by direct control of the remote mechanisms, or by higher level directives.

Teleoperator systems have high adaptability and low autonomy, due to the person being an integral part of the control loop. This works well for a certain class of problems. Robots tend to have high autonomy for specific, well-defined tasks, but low adaptability. We can get to autonomous systems by added intelligence and perception to robotic systems, or adding a supervisory mode to the telerobotic systems. NASA has sent robotic systems into Space since the 1970's. It could be argued that any spacecraft is a telerobot system, but we will take a narrower view.

Scope

This section covers where telerobotics are used and what they are used for.

In looking at the tasks that robots and telerobots are used for in the terrestrial environment, we gain insight into the tasks that can be used for in space.

Robots versus Humans, from an early USAF Study on military crews in space:

Humans	Robots
Complex tasks	well-defined tasks
Non-repetitive	repetitive
Life support	any environment
Rest & change	continuous operation
Dexterity	pre-planned tasks
Perceptual acuity	computation speed; communication speed
Decision making	consistency
Learning	if-then response
Intuition	limited contingency
Self-repair	
Replication	

In the nuclear industry, robots and telerobot systems are used to work in hazardous radiation environments. This certainly describes a lot of the environment outside the atmosphere of Earth. Some of the common tasks are materials handling and radiation monitoring. Similarly in the security and safety area, we use robotic systems for inspection and explosives handling. In HazMat situations, a robotic system can do reconnaissance and sampling. Similarly for fire fighting. Construction robotics for hazardous situations (such as mine cave-ins) are feasible.

Applications of telerobotics include:

Nuclear plant maintenance. Since the original Manhattan Project in the 1940's mechanical telerobot arms have been used to manipulate materials in high radiation environments. That's over 60 years of operational experience at this writing.

Toxic Chemical Handling. Telerobot systems are used in the handling (and, unfortunately, manufacture) of chemical munitions and nerve gas, and for cleaning up spills. Also, the application of handling and manipulation of hazardous biological materials.

Bomb disposal telerobots are widely used by the military and police forces around the world. The provide inspection and explosive ordnance neutralization without risking human operators. The German army was using wire-guided telerobotic systems in 1943.

Telerobot systems also find application in subsurface and underwater mining, particularly deep-water mining. They are used not only for exploration, but also for resource recovery (the Titanic, the Civil War Monitor, the Bismarck, and many others). They also find application in maintenance of underwater resources such as wells and communication cables.

Aerial Telerobot systems such as remotely piloted vehicles are deployed in many hostile areas, and have been watched with interest by domestic police forces. Military systems are now armed, and carry out lethal missions against designated targets. It is predicted that the current generation of fighter aircraft may be the last such manned vehicles produced.

Fire fighting autonomous and remotely operated systems provide service onboard ships and at airfields, and are preferred in toxic material fires. Many fire departments have units that can be deployed to reconnoiter hazmat spills by rail or highway vehicles.

Robots with a degree of autonomy have been used as warehouse or prison guards. Specialized units find use in pipeline maintenance, and decontamination, as well as surface cleaning. Remote controlled (telerobotic) trains and cranes are common.

Naturally, the use of telerobot systems has been extended to space, the ultimate in hazardous environments. The Mars rovers do not complain that they do not get to return home.

Prosthetics – arms, legs. In one sense, the artificial arms and legs are telerobotic systems, increasingly being controlled by nerve interfaces. In essence, the robot is part of you, not a remote unit.

Iron man suits and Exoskeletons.

A powered exoskeleton is a robot you wear. It is designed to augment, protect, and assist the wearer. It generally is responsive to the wearer's motion. We exclude for the moment the topic of medical prosthesis, discussed above. The major obstacle to the wider use of exoskeletons on Earth has been the power supply. Powered exoskeletons have been in use in research since the 1960's. Powered exoskeleton suits have been popular in science fiction, where the details of the power supply can be ignored. The military is funding much of the research in these areas. The applications in medical care areas is also being explored.

With a person inside the robot, there are some issues with joints and motion, as the axis of rotation of a joint (such as the hip) is outside the body with the suit, but inside the body for the wearer. Joint misalignment and slip result.

The advantage of making a telerobotic system with similar dimensions to a human is the ease of operation. For example, operation of an arm and manipulator with dimensions commensurate to a human's "feels" more natural to the operator.

Telerobotics can work for extended periods of time, using different operators in shifts. They can also "sleep" or continue in a low-power mode for extended periods before the mission begins - this is representative of planetary exploration systems during cruise phase.

Telerobots require less support infrastructure than humans.

In terms of evolution of capability, teleoperation systems can start out at 90% teleoperation/10% autonomy and grow or evolve to 10% teleoperation, 90% automation.

System components

Sensors

Sensors, or transducers, change their properties in response to an external event or condition. Usually we sense an external condition such as temperature, and translate that to a change in voltage, for example.

Sensors require signal conditioning electronics. This can involve amplification, filtering, conversion from analog, servo, or digital to another domain, and other such activities.

Non-contact sensors include proximity sensors that operate with inductive, capacitive, magnetic, or electrical/optical properties. Ranging and distance measurement involve sonar (in a fluid or atmosphere), radar or lidar tracking. Capacitive sensors detect by means of the dialectic properties of the target. Magnetic sensors, similarly, use the Target's magnetic properties. Light-based sensors can be passive, or active. Active sensors transmit or illuminate the target. Passive sensors work with ambient reflection. Vision is a complex, challenging option, requiring a lot of computation power, and clever algorithms. Some vision systems simplify the problem by using structured illumination or template matching. Shadowing is a problem, leading some autonomous land vehicles to stop for the shadow of a tree across a road. If there is one lesson that has been learned in decades of research and implementation in vision systems, it is that the brute force approach won't work. How humans process images is not well understood. The optic nerve, for example, has a channel capacity of about 5 bits per second. Evidently, a lot of processing is done at the sensor site.

Other issues in vision systems include dynamic range, perhaps enhanced by iris control. Color sensing both complicates and simplifies the problem. Most animals get along well without color vision. We can also choose to extend the frequency up into the ultraviolet or down into the infrared. The human vision system is self-cleaning, self-protecting, and to a certain extent self-repairing. Humans receive and process visual information on color, texture, and intensity. We derive information on three-dimensional shape, motion, and orientation. A key question is discrimination in a scene – what is the object, and what is the background? Boundaries may define an object, but they can also overlap. Depth perception, the third dimension, is tricky. If we know size, we can determine distance, and vice versa. Also, stereovision can use template matching for known objects. How could we use three eyes? No system in nature gives us a clue. Where do we put cameras? In a central location called "the head?" At the tip of each finger? How do we handle illumination? All these vision questions are beyond the scope of this discussion, but highly relevant.

For displacement measurement, we might use variable resistors encoders, or linear variable differential transformer (LVDT) sensors. Potentiometers, or variable resistors come in linear or rotary versions. The LVDT using an exciting current, which is picked up by sensor coils after being modified by the displacement of a ferromagnetic core. Rotary encoders use a precision optical disk to measure rotation angle, speed, or phase. The disc f the encoder, usually made of glass, has a precision pattern imprinted on it. It is read by a reflective or transmitive light sensor. As the disk rotates, a bit stream is generated. This can take the form of a Grey code, in which a single bit changes for each position change. Position determination depends on knowing the initial point, usually determined by a once per revolution index code. After that, it is a matter of counting pulses. Pulses per time give velocity. Looking at the leading edge of successive pulses can give very precise rotational position information.

Tactile or touch sensors can be built with printed inductors or capacitors on a flexible membrane. These are built in a similar manner to integrated circuits. They can be used to define a touch-map or surface profile.

What sensors do humans have? We might want to put these same sensors on a telerobotic device, and then extend the capabilities somewhat. Humans sense sound or vibration, light, with frequency discrimination (we call this color), pressure and force, slip, chemicals in aerosol or liquid form (smell, taste), temperature, and we monitor our own internal state. We might want to add environmental sensors such as radiation, chemical products, magnetic field, polarization, etc.

Chemical sensing is done in the human by taste (solution) and smell (aerosol). There are unfortunate cases, particularly involving some rocket fuels, where the human lethal limit is lower than the detection limit, as the author learned in his Launch Pad 39 orientation. We can device specific sensor for particular elements or compounds, or for a broad range of elements, such as hydrocarbons. We may want to sense pH, and we can go so far as to include a mass spectrometer to identify specific components of a sample. In robotics, we might use a *ChemFET*, or chemical field-effect transistor, which is a type of field-effect transistor acting as a chemical sensor. It is a structural analog of a MOSFET transistor, where the charge on the gate electrode is applied by a chemical process. It may be used to detect atoms, molecules, and ions in liquids and gases. It uses a gas-sensitive coating on the gate of the semiconductor. An *ISFET*, an ion-sensitive field-effect transistor, is a subtype of ChemFET devices. It is used to detect specific ions in electrolytes. An *ENFET* is a CHEMFET specialized for detection of specific biomolecules using enzyme reaction

What type of sensors would a robotic or telerobotic system use? We have our choice of off-the-shelf acoustic and electromagnetic sensors, tactile, force and slip sensors, temperature and humidity, specific chemicals, magnetic, position (using GPS), and internal state, such as battery state-of-charge, derived form an integration of input and output currents, and a good battery model. We might have particle sensors, such as smoke detectors, and a pH sensor. The manipulators might require angular and extension position in the many axes, and possible force sensing. Orientation sensors give us the angle with respect to local vertical (if we happen to be in a gravity field), or to magnetic north (if we are on a planet that has a significant magnetic field). We can also navigate and position with respect to an externally imposed grid, such as the terrestrial LORAN or GPS. We can also employ tip and tilt sensors. Drop-off sensors are useful for stairs and ravines. From touch sensors, we can build a touch map of a surface. Readings such as motor stall currents and temperatures of key components are useful.

Sensor data integration is a major topic in robotics. This includes not only sensor fusion – the merging of data from different sensor into a world-view, but sensor backup to resolve ambiguities. Our ability to collect bits exceeds our ability to process them. One solution is smarter sensors. Multiple sensors can be used with phase discrimination to determine time of arrival of a signal. We can do amplitude discrimination to get bearing information (we do this with our ears to locate the source of a sound). Sensor integration is the backing up of sensors with complementary sensors, to eliminate "holes' in the sensed fields. Sensors might be passive (listen only) or active, illuminating the target with specific energies.

Force feedback is a key feature for humans operating a telerobotic device. A graphical representation of the force is mostly useless. The operator should feel what the robotic system feels. This implies a back-drive capability at the controls. The bandwidth requirement is increased, but the force reflection becomes critical for effective control. Force scaling is preferred by operators to reduce fatigue.

Sensors in general are active or passive. Active sensors transmit a signal and look for a response. We can also determine the amount of data required to sense different conditions. For example, presence requires one bit. Intensity might be 4-16 bits. Change can be indicated by one bit, but amount or rate of change will require more. Distance, depending on the accuracy required might need 16 or more bits.

Actuators

Actuators provide manipulation and mobility to the system. A key metric is the number of degrees of freedom of the subsystem. The more degrees of freedom, the more complex the motion, and the more complex is the required control. Human tools are designed for humans to use. If we don't want to redesign tools, we design the robot manipulator to use human tools. The human arm has 2 degrees of freedom (DOF) in the shoulder, one at the elbow, and three at the wrist. The fingers provide more options. Geometrically, 6 degrees of freedom (3 rotation plus 3 translation) enable all possible motions. But the DOF pairs must be mutually exclusive for this to happen. Redundant DOF's enable alternate motions or paths for the same end point. In a task analysis, we determine the number of axes required. This translates to the minimum number of DOF's required. Our baseline is the human arm with 6.

A manipulator either holds a tool, or is the tool. It can be a screwdriver, or a dozer blade. It's were the work gets done. A gripper holds the tool, or the work piece. It can be a simple open-close clamp, a vacuum or magnetic grip, an analog of the human hand, or a wrapping device like a snake or tail.

An actuator is the drive mechanism (electric, pneumatic, hydraulic) that positions the manipulator. This includes the mobility system to move to the work site. The actuator payload is the load capacity at the gripper or end -effector, measured under static conditions. Dynamic conditions must also be considered. Accuracy usually refers to cyclic repeatability. We might consider accuracy as the minimization of error between expected (or commanded) state, and actual.

The end-effector might be a gripper arrangement, a tool, or a tool changer. The basic motions are linear and angular.

The sensor model for the human skin is one of high but varying sensitivity, highest near the fingertips. It has a fast response. It is flexible and durable, yet self-repairing. It is a smart sensor, containing a level of processing. Human skin can detect pressure, giving contour information; slippage, the pressure across a series of points; and temperature. This latter property allows for a level of materials identification by their thermal properties.

Actuators can have built-in embedded computers. These are referred to as smart actuators. They may incorporate a local feedback and monitoring loop. IEEE-1451 is a set of standards for interfacing smart sensors and smart actuators. The standards cover functions, communication protocols, and formats.

Electrically commutated motors

Electrically commutated, or brush-less direct current motors, are the latest variation on a technology that is more than a hundred years old. DC motors with brushes have been in commercial use since the 1880's, with demonstrations of the technology around 1840. They can produce large amounts of torque, but are inefficient, and high-maintenance because of the brush wear.

Brushless motors are a type of stepper motor, a configuration usually used for precise positioning. These are used to maintain specific rotation increments or rotary positions.

Brushless motors develop maximum torque at start, and have linearly decreasing torque at speed. They have permanent magnet rotors, and the armature windings are fixed. Electronics, perhaps an embedded controller, switches the phase of the motor windings to start and maintain the rotor rotation. This configuration leads to increased efficiency (more torque per watt, and more torque per weight), lack of brush wear and sparking, and less susceptibility to dirt and contamination.

Brushless motors require a position feedback, usually with a Hall effect or rotary encoder. The back-EMF induced in the un-driven coils may also be used to estimate position.

Methods of Navigation

Navigation in a structured environment is not a challenge. It's the real world that's a problem. A telerobot operates in a three-dimensional world, as do we. If they are not fixed, we need to measure a workspaces invariants, the dimensions, axes, possible navigation and reference points. A mobile robot can use dead reckoning or a beacon-based navigation. This works well on Earth, where we have navigation grid infrastructure such as GPS, but doesn't work on Mars. And Mars has no usable magnetic field. A key consideration for navigation is the choice of a coordinate system. For hundreds of years, Earth has had a well-defined latitude/longitude grid. There will be similar system for the Moon, and Mars. Navigation becomes obstacle avoidance when done up close.

Computation and Communication

Standard robotic systems use control systems such as servo or non-servo, point-to-point, or continuous. A non-servo system operates with two positions per path, the starting and ending point. A servo system can operate anywhere within its limits of motion. In a point-to-point control, the end points are specified, and the controller determines the path. In a continuous path approach, the robot is expected to smoothly follow the determined path. These are true for industrial painting robots and rovers on Mars.

In terms of degrees of freedom, the more DOF, the more complex the computational load for path planning, but more complex motions are allowed. Motions can be indexed, either breaking down a motion into way-points, or not moving all axes simultaneously. From the industrial world, most applications require three to five DOF for practical tasks.

Feedback is the key to stability in a system, an a necessary component of closed loop control. In a feedback loop, we compare the actual (position, velocity, etc) with the desired, and compute an error, a rate-of-change of error, etc. Decisions are based on sensory input and derived parameters.

Digital servo systems, and all but the most trivial are digital, are based on dedicated microcontrollers, or by shared control from a larger real-time central computer. There are trade-offs in computation power and power consumption, cooling, accuracy, etc. On the earliest Mars Rovers, it cost more energy to calculate a path for the vehicle to actually traverse that path.

A central processing unit can be a single computer that performs all control calculations. It is expensive, and must be fast. Dedicated microcontrollers at each axis can reduce the workload at the central computer. Sufficient computational power is the key to safety, and to the offloading of tasks from the human operator or adjunct.

Preferred processing units for space use at the moment include the RAD750 from BAE Systems, and the *Spacecube*, an FPGA-based approach. The latest architectures use a multicore approach.

The RAD750 is a radiation hardened single board computer based on a licensed version of the IBM PowerPC 750. The successor to the RAD6000, the RAD750 is manufactured by BAE Systems. It is intended for use in high radiation environments in space. The RAD750 was released for purchase in 2001 and the first units were launched into space in 2005. Software developed for the RAD6000 is upwardly compatible with the RAD750.

The cpu has 10.4 million transistors compared with the RAD6000's 1.1 million. It is manufactured using either 250 or 150 nm photolithography and has a die area of 130 mm² It has a core clock of 110 to 200 MHz and can process at 266 MIPS or more. The CPU can include an extended Level 2 cache to improve performance. Its packaging and logic functions are completely compatible with the standard PowerPC 750.

The CPU itself can withstand 2,000 to 10,000 gray (1 gray = 100 rad) and temperature ranges between –55 and 125C. It requires 5 watts. The standard RAD750 single-board system (CompactPCI board form factor) can withstand 1,000 gray and temperature ranges between –55°C and 70°C and requires 10 watts of power.

Maxwell's SCS-750 space computer incorporates three (Power-PC) PPC-750 chips in a voting configuration. The Mars Reconnaissance Orbiter, among many other spacecraft, uses the RAD-750. It is estimated that there were some 200 units on space missions by 2000.

Sensor Interfacing

Sensor interfaces to the computer mostly consist of analog to digital conversion, with some direct digital and timing circuits. What is the user interface? For a robot it might be a simple command link via hardwire or radio. This is for operations; in the lab, a pc would be used for development, simulation, debugging, and download.

Whether a single or multiple computers, it is a classic embedded computer system. This implies limited or no human interface, and a multitasking, real-time program.

Simpler systems can get by with a simple sequential task loop, but rapidly reach the complexity point where a priority-driven task list is necessary. This is complicated by multiple asynchronous interrupts form external devices requiring service.

Tasks are work units that the system must accomplish. These are mapped into software entities (also called tasks) which have an associated priority. At the top level, we must carefully define what has to be done, how often, how fast, and the relative order of importance, which might change as a result of state. For example, in an emergency, priorities change.

Individual axes or actuators might have their own processors, a dedicated processor that is adequate for the well-defined role. They have specific interfacing requirements, and implement a particular control law or set of restraints that are "hard-wired" into the units. These may change as a result of downloading from the higher-level controller. For example, the controller for a particular mechanism might be reprogrammed in response to a sensor or mechanical failure. Output is usually a drive signal to a power interface to a mechanism. This might involve digital to analog conversion, servo conversion, scaling, or other special data processing. We can command position, velocity, acceleration, force, torque, or other quantities, that the actuator then realizes for us. We use feedback via sensors to determine if the desired result was correctly implemented.

Moving up the ladder of complexity, a more capable processor might tackle forward and inverse kinematic transformations, motion planning, and world modeling.

The coupling between processors between layers can be loose or tight. If we think of the processors as networked, the coupling between units determines the autonomy that lower level units have. We might have an aristocratic hierarchy, or a democracy. Loose coupling is characterized by modest communication speeds and fairly simple, predetermined coordination. Sometimes, loosely coupled systems are too slow, and the communications is complicated. Tight coupling implies that we have direct data access between processors, and fast communication, usually with a reliable handshake mechanism. This might be the case for multiple co-operating systems on the same bus. In tightly coupled system, it can be a problem to debug the communication links. Bus contention can be an issue, and processor arbitration is a difficult tasks at high speeds.

Sensor processor's offload the computation processor by doing sensor data calculations. They are usually considered a part of the sensor. They take the raw sensor data and provide it in a processed package that is "easier to digest."

A hierarchical model for robot control, similar to the OSI model for communication,s was developed at the National Bureau of Standards (NBS), later, National Institute of Standards and Technology (NIST) in Gaithersburg, Maryland. Initially addressing factory automation, the model has proven to be very versatile in addressing a wide variety of tasks.

The model describes a series of functions, and the data and control flow between the functional layers. From the bottom up, the typical layers for a hierarchical robot/telerobot control systems would be sensor input, servo, primitive command, kinematics, path planner, command exec, and user level. With well-defined interfaces between each level, development can proceed in parallel, and various levels can be generic.

At the user level, we would define robot positions and motion sequences, and storage and retrieval of information. We might go so far as to define a robot control language to use at this level. Storage and retrieval of information involves programs, associated data, descriptions, and task-unique information.

At the command exec level, the basic functions of the robot are implemented. This level receives higher-level commands from the user level, and generates specific commands for the path planner level.

At the path planner level, motions are decomposed into sub-motions. For example, at the command exec level, we may want the robot to move to position X. At the path planner level, we decide how that motion is going to be executed. It depends on whether we walk, roll, swim, or fly. But the next level up does not need to care.

At the kinematics level, the commands form the path planner are received, the inverse kinematic transform is solved, and commands are issued to the primitive command level.

The primitive command level receives commands from the kinematics level, and implements these, subject to specific joint-related constraints. This level issues commands to the actuator servo level. If we change actual mechanisms, this level need not change, but the next level down will.

The actuator servo level receives commands from the primitive command level, and sends the commands to the appropriate actuator electronics. A translation in signal level or properties might be required. Multiple actuators can be controlled.

Sensory information processing pre-processes any environmental inputs, receives commands from any level, and sends processed sensory data to the appropriate level.

The actuator control systems can be fairly simple structures, such as position and rate-based control systems. Rate information can be problematical, but does enhance stability and improve transient response. Rate can be measured directly, or derived from repeated position measurements in software or hardware.

Software for robotic systems will be classical embedded real-time systems. They must be sensory-interactive. The actual choice of language doesn't matter, as long as it supports real-time concepts. Some languages are better than others at this, and ease of debugging is a major factor. As with any big important software project, complete and correct requirements documentation are critical. From these requirements, a flow-down of verifiable specifications must be derived. Then the usual steps of prototyping, implementation, and testing are accomplished. It has been shown numerous times in real-time embedded systems that the more effort spent of the initial steps, the easier the later steps (i.e., debug) will be.

A good simulation goes a long way in verifying the system. It can be used to verify assumptions, verify strategy and algorithms, check timing constraints, and extract program metrics. Of course, the simulation itself must be verified complete and correct.

Early spaceflight computers were custom designs, but cost and performance issues have driven the development of variants of commercial chips. Aerospace applications are usually classic embedded applications. Space applications are rather limited in number, and, until recently, almost exclusively meant National Aeronautics and Space Administration (NASA),

European Space Association (ESA), National Aero Space Development Agency (NASDA) - Japan, or some other government agency. Flight systems electronics usually require MIL-STD-883b, Class-S, radiation-hard (total dose), SEU-tolerant parts. MIL-STD-883 is the standard for testing and screening of parts. Specific issues of radiation tolerance are discussed in MIL-M-38510. Class-S parts are specifically for space-flight use. Because of the need for qualifying the parts for space, the state-of-the-art in spaceborne electronics usually lags that of the terrestrial commercial parts by 5 years.

The advantage of a radiation-hard version of a commercial chipset is the ability to piggyback onto the existing set of software tools, applications, and development environments, as well as a pool of software developers.

Processors used in aerospace applications, as any semiconductor-based electronics, need to meet stringent selection, screening, packaging and testing requirements, and characterizations because of the unique environment. Most aerospace electronics, and the whole understanding of radiation effects, were driven by the cold war defense buildup from the 1960's through the 1980's. This era was characterized by the function-at-any-cost, melt-before-fail design philosophy. In the 1990, the byword was COTS -- use of Commercial, Off-The-Shelf products. Thus, instead of custom, proprietary processor architecture's, we are now seeing the production of specialized products derived from commercial lines. In the era of decreasing markets, the cost of entry, and of maintaining presence in this tiny market niche, are prohibitively high for many companies.

FPGA -based solutions are now mainstream for spacecraft computing usage, using hard or soft-cores of standard microprocessor architectures. Either the entire structure is constructed to be radiation-hard, or triplication with selected hardened circuits is applied.

This section discusses the major environmental obstacles to the use of sophisticated electronics in space, and the mitigation techniques that can be applied.

Radiation

Here on the surface of the planet, we are mostly shielded by the atmosphere and by the magnetic field lines (but only against charged particles) from the fury of the Sun. In space, or on other planetary surfaces, the situation is much different.

There are two radiation problem areas: cumulative dose, and single event. Operating above the Van Allen belts of particles trapped in Earth's magnetic flux lines, spacecraft are exposed to the full fury of the Universe. Earth's magnetic poles do not align with the rotational poles, so the inner Van Allen belts dip to around 200 kilometers in the South Atlantic, leaving a region called the South Atlantic Anomaly. The magnetic field lines are good at deflecting charged particles, but mostly useless against electromagnetic radiation and uncharged particles such as neutrons. One trip across the Van Allen belts can ruin a spacecraft's electronics. Some spacecraft turn off sensitive electronics for several minutes every ninety minutes – every pass through the low dipping belts in the South Atlantic.

The Earth and other planets are constantly immersed in the solar wind, a flow of hot plasma emitted by the Sun in all directions, a result of the two-million-degree heat of the Sun's outermost layer, the Corona. The solar wind usually reaches Earth with a velocity around 400 km/s, with a density around 5 ions/cm^3. During magnetic storms on the Sun, flows can be several times faster, and stronger. The Sun has an eleven year cycle of maxima. A solar flare is a large explosion in the Sun's atmosphere that can release as much as 6 × 10^{25} joules in one event, equal to about one sixth of the Sun's total energy output every second. Solar flares are frequently coincident with sun spots. Solar flares, being releases of large amounts of energy, can trigger Coronal Mass Ejections, and accelerate lighter particles to near the speed of light.

The size of the Van Allen Belts shrink and expand in response to the Solar Wind. The wind is made up of particles, electrons up to 10 Million electron volts (MeV), and protons up to 100 Mev – all ionizing doses. One charged particle can knock thousands of electrons loose from the semiconductor lattice, causing noise, spikes, and current surges. Since memory elements are capacitors, they can be damaged or discharged, essentially changing state.

Not that just current electronics are vulnerable. The Great Auroral Exhibition of 1859 interacted with the then-extant telegraph lines acting as antennae, such that batteries were not needed for the telegraph apparatus to operate for hours at a time. Some telegraph systems were set on fire. The whole show is referred to as the Carrington Event, after British Scientist Richard Carrington, who observed the event.

Around other planets, the closer we get to the Sun, the bigger the impact of solar generated particles, and the less predictable they are. Auroras have been observed on Venus, in spite of the planet not having an observed magnetic field. The impact of the solar particles becomes less of a problem with the outer planets. Auroras have also been observed on Mars, and the magnetic filed of Jupiter, Saturn, and some of the moons cause their "Van Allen belts" to trap large numbers of energetic particles, which cause more problems for spacecraft in transit. Both Jupiter and Saturn have magnetic field greater than Earth's. Not all planets have a magnetic field, so not all get charged particle belts.

Radiation Hardness Issues for Space Flight Applications

A complete discussion of the physics of radiation damage to semiconductors is beyond the scope of this document. However, an overview of the subject is presented. The tolerance of semiconductor devices to radiation must be examined in the light of their damage susceptibility. The problems fall into two broad categories, those caused by cumulative dose, and those transient events caused by asynchronous very energetic particles, such as those experienced during a period of intense solar flare activity. The unit of absorbed dose of radiation is the rad, representing the absorption of 100 ergs of energy per gram of material. A kilo-rad is one thousand rads. At 10k rad, death in humans is almost instantaneous. One hundred kilo-rad is typical in the vicinity of Jupiter's radiation belts. Ten to twenty kilo-rad is typical for spacecraft in low Earth orbit, but the number depends on how much time the spacecraft spends outside the Van Allen belts, which act as a shield by trapping energetic particles.

Absorbed radiation can cause temporary or permanent changes in the material. Usually, neutrons, being uncharged, do minimal damage, but energetic protons and electrons cause lattice or ionization damage in the material, and resultant parametric changes. For example, the leakage current can increase, or bit states can change. Certain technologies and manufacturing processes are known to produce devices that are less susceptible to damage than others.

Radiation tolerance of 100 kilo-rad is usually more than adequate for low Earth orbit (LEO) missions that spend most of their life below the shielding of the Van Allen belts. For Polar missions, a higher total dose is expected, from 100k to 1 mega-rad per year. For synchronous, equatorial orbits, that are used by many communication satellites, and some weather satellites, the expected dose is several kilo-rad per year. Finally, for planetary missions to Venus, Mars, Jupiter, Saturn, and beyond, requirements that are even more stringent must be met. For one thing, the missions usually are unique, and the cost of failure is high. For missions towards the sun, the higher fluence of solar radiation must be taken into account. The larger outer planets, such as Jupiter and Saturn, have large radiation belts around them as well.

Cumulative radiation dose causes a charge trapping in the oxide layers, which manifests as a parametric change in the devices. Total dose effects may be a function of the dose rate, and annealing of the device may occur, especially at elevated temperatures. Annealing refers to the self-healing of radiation induced defects. This can take minutes to months, and is not applicable for lattice damage. The total dose susceptibility of the Transputer has been measured at 35-50 k-rad with no internal memory. The internal memory or registers are the most susceptible area of the chip, and is usually deactivated for operations in a radiation environment. The gross indication of radiation damage is the increased power consumption of the device, and one researcher reported a doubling of the power consumption at failure. In addition, failed devices could operate at a lower clock rate, leading to speculation that a key timing parameter was being effected in this case.

Single event upsets (seu's) are the response of the device to direct high energy isotropic flux, such as cosmic rays, or the secondary effects of high energy particles colliding with other matter (such as shielding). Large transient currents may result, causing changes in logic state (bit flips), unforeseen operation, device latch-up, or burnout. The transient currents can be monitored as an indicator of the onset of SEU problems. After SEU, the results on the operation of the processor are somewhat unpredictable. Mitigation of problems caused by SEU's involves self-test, memory scrubbing, and forced resets.

The LET (linear energy transfer) is a measure of the incoming particles' delivery of ionizing energy to the device. Latch-up refers to the inadvertent operation of a parasitic SCR (silicon control rectifier), triggered by ionizing radiation. In the area of latch-up, the chip can be made inherently hard due to use of the Epitaxial process for fabrication of the base layer. Even the use of an Epitaxial layer does not guarantee complete freedom from latch-up, however. The next step generally involves a silicon on insulator (SOI) or Silicon on Sapphire (SOS) approach, where the substrate is totally insulated, and latch-ups are not possible.

In some cases, shielding is effective, because even a few millimeters of aluminum can stop electrons and protons. However, with highly energetic or massive particles (such as alpha particles, helium nuclei), shielding can be counter-productive. When the atoms in the shielding are hit by an energetic particle, a cascade of lower energy, lower mass particles results. These can cause as much or more damage than the original source particle.

Mitigation Techniques

The effects of radiation on silicon circuits can be mitigated by redundancy, the use of specifically radiation hardened parts, Error Detection and Correction (EDAC) circuitry, and scrubbing techniques. Hardened chips are produced on special insulating substrates such as Sapphire. Bipolar technology chips can withstand radiation better than CMOS technology chips, at the cost of greatly increased power consumption. Shielding techniques are also applied. In error detection and correction techniques, special encoding of the stored information provides a protection against flipped bits, at the cost of additional bits to store. Redundancy can also be applied at the device or box level, with the popular Triple Modular Redundancy (TMR) technique triplicating everything, and based on the assumption that the probability of a double failure is less than that of a single failure. Watchdog timers are used to reset systems unless they are themselves reset by the software. Of course, the watchdog timer circuitry is also susceptible to damage.

Thermal issues

Radiation is not the only problem. In space, things are either too hot or too cold. On the inner planets toward the Sun, things are too hot. On the planets outward of Earth, things are too cold. In space, there is no gravity, so there are no conduction currents. Cooling is by conduction and radiation only. This requires heat-generating electronics to have a conductive path to a radiator.

That makes board design for chips, and chip packaging, complex and expensive.

Mechanical issues

In zero gravity, every thing floats, whether you want it to or not. Floating conductive particles, bits of solder or bonding wire, can short out circuitry. This is mitigated by conformal coatings, but the perimeter of the chip die is usually ground, and cannot be coated due to manufacturing sequences.

The challenges of electronics in space are daunting, but much is now understood about the failure mechanisms, and techniques to address them.

Spaceflight processors

The RAD750 is a radiation hardened single board computer based on a licensed version of the IBM PowerPC 750. The successor to the RAD6000, the RAD750 is manufactured by BAE Systems. It is intended for use in high radiation environments in space. The RAD750 was released for purchase in 2001 and the first units were launched into space in 2005. Software developed for the RAD6000 is upwardly compatible with the RAD750.

The cpu has 10.4 million transistors compared with the earlier RAD6000's 1.1 million. It is manufactured using either 250 or 150 nm photolithography and has a die area of 130 mm² It has a core clock of 110 to 200 MHz and can process at 266 MIPS or more. The CPU can include an extended Level 2 cache to improve performance. Its packaging and logic functions are completely compatible with the standard PowerPC 750.

The CPU itself can withstand to 1 megaRAD and temperature ranges between –55 and 125C. It requires 5 watts. The standard RAD750 single-board system (CompactPCI board form factor requires 10 watts of power.

Maxwell's SCS-750 space computer incorporates three PPC750 chips in a voting configuration. By 2010, it was estimated that there were over 200 RAD750s used in a variety of spacecraft.

Freescale Semiconductor manufacturers a series of PowerPC-based embedded microcontrollers. These have evolved to include multicore models. In particular, the 8-core P4080 was chosen by the Fraunhofer Institute for Computer Architecture and Software Technology for their Project Muse. Muse stands for Multicore architecture for sensor-based position tracking in Space. The chosen chip can operate with up to a 1.5 GHz clock, to achieve a processing power of 60 GIPS. The chip is built on silicon-on-insulator technology for radiation tolerance. Multiple cores are used for redundancy and fault tolerance. The TMR voting circuitry is implemented in a radiation tolerant FPGA. The microcontroller contains, besides the 8 computer cores, six gigabit ethernet channels, dual PCI express interfaces, dual Rapid I/O, and dual SpaceWire.

An Architectural Model

This section presents an architectural model for robotic systems, developed at the National Bureau of Standards (later, National Bureau of Standards and Engineering).

NASREM

The NASA/NBS Standard Reference Model for Telerobot Control System Architecture was evolved as a model for the implementation of advanced control architectures.

The NBS architecture is a generic framework in which to implement intelligence of a telerobotic device. It was developed over a decade as part of a research program in industrial robotics at NBS (now. NIST) in which over $25 million was spent. The NBS program involved over fifty professionals and extensive facilities, including robots, a supercomputer, mainframes. minicomputers. microcomputers. LISP machines. and AI workstations. This model, designed originally for industrial robots. is the mechanism by which sensors. expert systems. and controls are linked and operated such that a system behaves with some measure of autonomy, if not intelligence.

Systems designed from this model perform complex real-time tasks in the presence of sensory input from a variety of sensors. They decomposes high level goals into low level actions. making real-time decisions in the presence of noise and conflicting demands on resources. The model provides a framework for linking artificial intelligence. expert system. and neural techniques

28

with classical real-time control. Sensors are interfaced to controls through a hierarchically-structured real-time world model. The world model integrates current sensory data with a priori knowledge to provide the control system with a current best estimate of the state of the system.

NASREM is a generic hierarchical structured functional model for the overall system. The hierarchical nature makes it ideal for telerobot systems, and for gradual evolution of the system. The model also provides a set of common reference terminology, which can enable the construction of a database. It defines interfaces, which allows for modularization. The model allows for evolutionary growth, while providing a structure of the interleaving of human:robotic control.

NASREM's 6-level model operates from a global memory (or database). At each level we have three processes, sensory processing world modeling, and task decomposition (execute). At the very lowest level, we have the raw sensors and the servo systems. Going up from that, we have the primitive level, the elementary move level, the task level, the service bay level, and the mission level. At the servo level, we would find cameras, and their associated pan/tilt control as well as mobility and joint motor control, with associated position feedback. At the primitive move level, we would find the camera subsystem, the arm, the mobility subsystem, and the grippers. At the elementary (or e-) move level, we would find systems such as perception or manipulation. At the task level, we might locate the entire telerobotic system.

The world modeling process starts with a sparse database. Sensor data, appropriate to the level flows in, and there might be a capability for data fusion. A task planner task can make "what-if" queries of the world model (which is state-based). The modeling task uses a global database of state variable, lists, maps and knowledge bases to allow a modeling process to update and predict states, to evaluate current states and possible states, and to report results to a task executor task. The World model, evaluates states, both existing states as evidenced by sensor data, and possible states, as postulated by the task planner.

The timing and time horizon of the various levels of the model is are vastly different. The servo level operates on the millisecond level, the primitive level, at 10's to 100's of milliseconds, and the e-move level at about a one second update interval. It would have about a 30 second planning horizon.

The task level would have update interval on the order of seconds to 10's of seconds, with a planning horizon in the 10's of seconds. Moving up, the service by level would update in seconds, with a planning horizon the order of minutes to 10's of minutes. Finally, the mission level might update on the order of minutes, with a horizon of an hour.

The servo level would accept Cartesian trajectory points from the next level up, and transform these to drive voltages or current for the mechanisms. The Primitive level would accept pose (or collection of joint angles and positions) information from the next higher level, and generate the Cartesian trajectory point to pass down the hierarchy. These involve dynamics calculations. The e-move level would accept elementary move commands and generate pose commands, after orientations in the coordinate frame, singularities, and clearances. It uses simple if-then state transition rules. The task level, the one the telerobot would be located at, accepts task commands (from the human operator), does subsystem assignments and scheduling, and generates a series of e-moves.

Real Time Control System (RCS)

RCS evolved from NASREM over decades, starting in the 1970's. RCS is a Reference Model Architecture for real-time control. It provides a framework for implementation in terms of a hierarchical control model derived from best theory and best practices. RCS was heavily influenced by the understanding of the biological cerebellum. NIST maintains a library of RCS software listings, scripts and tools, in ADA, Java, and C++.

An abstraction, the perfect joint accepts analog or digital torque commands, and produces the required torque via a dc motor. It also provides state feedback in the form of force, torque, angle or position, (depending on whether the joint configuration is Cartesian or revolute, and possibly rate. The perfect joint includes a pulse width modulator (pwm), a motor, and possibly a gearbox. Internal feedback and compensation is provided to compensate for gearbox or other irregularities such as hysteresis or stiction, For example, the torque pulses common to harmonic drives can be compensated for within the perfect joint. The perfect joint is part of the lowest NASREM level. The processing provided theoretically achieves a "perfect" torque, where the outputted torque matches the commanded torque.

Safety and Human Factors

Robotic and Telerobotic systems, particularly those designed to operate in conjunction with humans, need inherent designed-in safety. Like for anything mechanical, we need to do a human factors analysis of the system. Other industry-standard hazardous analyses such as Failure Modes & Effects Analysis (FMEA), fault tree, sneak path, etc. are required. The robot system must be aware of the intrusion of a human into its workspace. In flight systems, safety is a significant complexity and cost driver. Robotic safety has a basis in industrial applications. Flight safety for systems operated in proximity with humans has a heritage in the Space Shuttle and Space Station programs. There is a scarcity of data on robots in Zero-g and in proximity with astronauts, but this is changing with the Robotnaut, now on the International Space Station. What is certainly true is that safety has to be designed-in from the start, not added on later.

Years ago, Science fiction author Isaac Asimov introduced his famous "Three laws of Robotics":

•A robot may not injure a human being, or through inaction, allow a human being to come to harm.

•A robot must obey the orders given it by human beings except where such orders would conflict with the First law.

•A robot must protect its own existence as long as such protection does not conflict with the First or Second Laws

Of course, a robot must be aware of a human to apply the laws. Although these are still good guidelines, top-level requirements, their application in a robotic system always presents challenges.

Robotic systems have been in use for over fifty years, and there have been lessons learned. Based on a large number of training accidents, there is an increased need for simulation and scenario planning in an offline system. Accidents during maintenance have shown that the human in the work envelope of the robot must have complete and total power control, and a method of verifying this. The robotic system must have a failsafe and verifiable disable or safe mode.

When one human works next to another in a hazardous environment, there are certain reasonableness assumptions made. Humans anticipate reach, start of motion, speed, etc. Thus, there is in a sense the need to "make robots in our image", so our built-in assumption base works. This has been referred to as anthropomorphic chauvinism, where we assume the remote unit has bilateral symmetry, binocular vision, kinematics like the human body, and human performance bounds.

Human contact by a robot system must be precluded by design. This is a corollary of Asimov's Rule 1, but is not easy to implement. Contact must also be prevented by operational procedure (verified by simulation). The robot needs to be made aware of the human presence. Accidental release of tools or workpieces must be precluded by design, using tethers, constraints, or other means. This case must also be verified by simulation of the operational procedures.

It goes without saying, but I'll say it anyway: The simulation itself must be certified and verified to represent the actual robotic system and the workplace – otherwise it is useless at best, and dangerous at worst.

In the history of the use of robots in proximate locations with humans, what have been the causes of accidents? Contact during teaching or maintenance, inadvertent release of a tool or workpiece, and unexpected or unpredicted motion. The latter is caused by control error, mechanical error or failure, or procedural error – all preventable.

Robots have a better safety record in industry that other industrial equipment, and have caused fewer fatalities. Is it because we humans are more wary of these systems?

We need to prevent, by design, human-robot contact. The human will avoid the robot if he/she can. The robot needs to know the human is present. Human performance modeling as part of the robot's world knowledge is essential.

Human factors issues include data input to the human in terms of voice, video, digital display, force reflection and audible alerts. The volume of input data must be prioritized and controlled to prevent information overload. The control output is produced by the human by means of switches, voice command, joysticks and trackballs, a keyboard (bad choice), or replica master systems, that mimic the movement of human hands or fingers.

We also need to consider the human's workload in terms of physical and cognitive fatigue.

When we put a human into a hazardous environment such as space in a protective enclosure, we need to consider factors such as heat and moisture removal, odor control, the dexterity, mobility, and sensory attenuation, the drop in efficiency, the shift in the center of gravity with a suit with a backpack in a gravity field, and the shift in body image perception. The suited human is bigger and bulkier than usual..

If we have a human operator at a telerobotic control workstation, we need to carefully consider how to display data, and accept control input. As mentioned, we can use a replica master system, but that assumes the robotic manipulator is a close kinematic copy of the human arm. Force reflection is also used in these devices to backdrive the human end in proportion to the sensed force at the robot end. The force can be scaled to be a fraction of that sensed, to reduce fatigue.

There are numerous standards for human factors issues. NASA has quite a few. These define safety issues, and such parameters as visual cone and reach envelope, and design of the human:machine interface.

Some key design issues include the minimization of operator workload by need for simultaneous control of disparate systems, and the design of graphical overlays for visual systems, giving additional data on position, velocity and torques as well as wire-frame graphical models superimposed on the scene. The view might be presented in hand controller coordinates. Text can be overlaid for status and prompts. This merging of sensed and derived data is computationally intensive. Ergonomic considerations require us to accommodate a wide range of operators. The human workstation should be "task-transparent."

The many aspects of space robotic systems bring unique safety requirements. The robotic system cannot endanger the mission, the launch vehicle, the human crew, or other equipment. Space robotics projects are high visibility, with little or no room for failure.

The Flight Telerobotic System was to operate in close proximity to the Space shuttle or other launch vehicles, to the space Station, to other spacecraft, and to EVA astronauts. The robotic element may operate within the habitable volume of the Space Station. It may operate alone, or in proximity to EVA astronauts. In such cases, the robotic element can extend the envelope of human capability, and may relieve safety constraints, imposed to protect human life.

In the context of the control model, perhaps NASREM-derived, an anomaly must be sensed, and then becomes the subject of a contingency replanning task.

Swarms

This section describes a different approach to robotics: collections of smaller co-operating multi-robotic systems that can combine their efforts and work as ad-hoc teams on problems of interest.

This is based on the collective or parallel behavior of homogeneous systems. This covers collective behavior, modeled on biological systems. Examples in nature include migrating birds, schooling fish, and herding sheep. A collective behavior emerges form interactions between members of the swarm, and the environment.

A driver in the space environment is the exploration of the asteroids, numbering in the thousands. Although there are fewer than 10 planets, and less than 200 moons, there are millions of asteroids, mostly in the inner solar system. The main asteroid belt is between Mars and Jupiter. Each may be unique, and some may provide needed raw materials for Earth's use. There are three main classifications: carbon-rich, stony, and metallic.

The physical composition of asteroids is varied and poorly understood. Ceres appears to be composed of a rocky core covered by an icy mantle, whereas Vesta may have a nickel-iron core. Hygiea appears to have a uniformly primitive composition of carbonaceous chondrite. Many of the smaller asteroids are piles of rubble held together loosely by gravity. Some have moons themselves, or are co-orbiting binary asteroids. The bottom line is, asteroids are diverse.

It has been suggested that asteroids might be used as a source of materials that may be rare or exhausted on earth (asteroid mining) or materials for constructing space habitats or as refueling stations for missions. Materials that are heavy and expensive to launch from earth may someday be mined from asteroids and used for space manufacturing. Valuable materials such as platinum may be returned to Earth for a profit.

Exploring the asteroids requires a diverse and agile system. Thus, a swarm of robotic spacecraft with different capabilities might be used, combining into Teams of Convenience to address situations and issues discovered in situ.

In Swarm robots, the key issues are communication between units, and cooperative behavior. The capability of individual units does not much matter; it is the strength in numbers. Ants and other social insects such as termites, wasps, and bees, are models for robot swarm behavior. Self-organizing behavior emerges from decentralized systems that interact with members of the group, and the environment. Swarm intelligence is an emerging field, and swarm robotics is in its infancy.

Projects – Case Studys

In this section, we will examine some space robotics/telerobotics systems.

Lessons Learned from Underwater Systems

In shallower water, human divers and telerobotic systems, operated from the surface, can work in proximity. In special cases, human divers can operate at 1,000 feet, accompanied by observation-class robotic systems. At greater depths, telerobotic systems are on their own, but several systems, perhaps controlled by different operators, can cooperate on tasks.

These has been a commercial industry in the area for many years. Major uses include the offshore oil industry. The observation vehicles are basically sensor platforms, with a frame, flotation, and thrusters. They can be tethered or free-swimming. The work class machines also include cameras (and associated lighting), manipulators, perhaps with interchangeable tooling, and flotation and thrusters.

Typical tasks for the telerobotic systems include clearing or cutting tangled hoses, site survey, inspection, placing and recovering acoustic beacons, clearing debris, recovering dropped equipment, and placing equipment packages.

Flight Telerobotic Servicer

The Flight Telerobotic Servicer was a NASA Project circa 1987. It was to develop "a safe, reliable, and useful tool for Space Station Freedom assembly, maintenance, servicing, and inspection tasks." It was, for a while, the focus of NASA's Automation and Robotics Development efforts. Johnson Space Center took the lead in defining the program and requirements. The project was the result of a Congressional Mandate to include more Robotics and Automation in the Space Station Program. It had a goal of reducing on-orbit labor requirements.

The project involved the development of a robotic unit used externally on the station, with teleoperator control from within the pressurized space. It was to support crew activities during assembly, maintenance, servicing and inspection. It would be the choice for repetitive or hazardous tasks. Autonomy was a major cost-driver, but a graduation evolution to autonomy of the system was seen. Units and assemblies could be designed to be "robot friendly" from the onset, with structured environments and databases of parts, assemblies, and ORU's – orbital replacement units. The telerobot itself was an (Orbital Replacable Unit) ORU. Extensive consideration was given to built-in self test, diagnostics, and ease of repair. It was to be failsafe/fail operational with redundancy, and manual intervention as an override. The tooling at the end of the arms was to be easily swapped out.

Working in space in a space suit is tiring and inefficient. First, there is a pre-breathing period, since the suit and the Station do not use the same gas mixtures. The suits are bulky, because they carry with them a complete environment conducive to life. Visibility is limited, dexterity is constrained, temperature and humidity control is difficult, and taking a bathroom break is not feasible. This is a similar situation to deep-sea divers, although in that case, the delay comes at the ascent, when stops must be made to minimize "the bends."

The FTS was designed to be capable of replacing ORU's. These are assembly's that are specifically designed to be changed out on-orbit, similar to the concept of line replaceable units. The FTS was to be stored outside the habitable volume of the Space Station, in vacuum. It would be powered from the Station at that time, and minimally awake, capable of running self-testing diagnostics. It would mount both cameras and a lighting system. The arm segments we to be mechanically and electrically standardized, and repairability was enhanced by the use of ORU's for the telerobot itself. The arm modules would use smart appendages.

FTS was to operate anywhere on the station where an appropriate attach point was provided, or from the Shuttle. The FTS was to be compatible with the RMS, the Shuttle's arm. It could be operated in telerobotic mode from the shuttle or the Station itself. On the Station, the FTS would be positioned at a work site by the Space Station's mobile arm/crane. The FTS's arms would have 6 degrees of freedom, 3 rotary and 3 prismatic. This provided a 8 foot radius workspace sphere. Power requirements were estimated to be 580 watts peak, 360 watts average, 100 watts in sleep-storage mode. A hold-up battery in the unit would allow safe shutdown in case of power disruption. The unit was to weigh about 1800 pounds. There would be two global cameras, and 2 wrist cameras. The FTS interface to payloads would be the same as the Shuttle RMS's - a standard payload grapple fixture. the FTS would be transported to locations on the station by a separate mobility system using both permanent and portable rails.

For the initial assembly of the Station on-orbit, the FTS was to be capable of installing and removing trusses and ORU's, and mating various connectors. It would also be used in inspection tasks, and to mate thermal utility connectors.

A major goal of the program was how to test the robot assembly in the gravity field. For the shuttle RMS, an air-bearing floor allows 2-D testing. The arm itself cannot support its own weight in 1-G and, the dynamic control is not the same in a gravity field.

After the assembly of the Station was complete, the FTS was expected to complete its task autonomously. An initial plan for the Station was to provide an on-orbit servicing facility for spacecraft. The FTS was a key component of this plan, being designed to replace ORU's on the Hubble Space Telescope, or to refuel expendables in selected spacecraft.

The FTS study's assumed an indefinite on-orbit lifetime with periodic maintenance. It was to be capable of operating up to 30 hours per week. It was going to be a 1,500 pound device, requiring about a kilowatt of power.

At the time, the State-of-the-art in Spacecraft computers was a radiation-hard version of the Intel 80386 chip. The software architecture was based on the NIST/NASA NASREM model. The FTS was designed to be operated by a human, in supervised autonomy mode, or completely autonomously, for selected tasks. It was designed to be modular and upgradeable, so the design could evolve. The evolution of AI and expert systems was expected to make the FTS more capable as time went on, and processing, storage, and communications would evolve. 1-10 MIPS of processing power was assumed available for the system. Flight systems necessarily lag the state-of-the-art in terrestrial, desktop or embedded systems due to the unique environmental considerations. The FTS had a requirement of controlling two seven-degree of freedom manipulators with this level of processing throughput.

An example of a supervised autonomy operation would be the tightening of a bolt. The human in the loop would locate the bolt, using his or her superior vision processing. When properly positioned, the robot system would tighten the bolt. As machine vision systems evolved, the robot would be able to do the location task autonomously.

Safety would be provided by momentum and velocity limiting, running lights, and an independent watchdog computer system with the authority to remove power.

Sufficient embedded computational power is the key to safety, offloading of mundane, repetitive tasks from the human operator, and transition to autonomous operations. Planned designs for the FTS telerobot included 1-10 MIPS of processing power. The evolutionary system would require 1-2 orders of magnitude more computational power, but this appeared to be both achievable and affordable within the serviceable lifetime of the planned system.

Several architectures existed that allowed parallel processing with existing and emerging cpu's that were applicable to an evolutionary system. The system requirements for on-orbit servicing by teleoperators were mapped into the proposed architectures. The special computational and communication needs of safety were covered. This design was in accordance with the NASREM architecture and met the system safety requirements of NHB-1 700-7B.

In the FTS design, the key requirements were the telerobotic control of multiple degrees of freedom manipulators. The resulting computational requirements were not easily met by existing hardware, and strained the limits of flight qualified systems in the near term. Several abstractions were used to decompose the system design into manageable units.

An abstraction, the perfect joint accepts analog or digital torque commands, and produces the required torque via a dc motor. It also provides state feedback in the form of force, torque, angle or position, (depending on whether the joint configuration is Cartesian or revolute), and possibly rate. The perfect joint includes a pulse width modulator (pwm), a motor, and possibly a gearbox. Internal feedback and compensation is provided to compensate for gearbox or other irregularities such as hysteresis or stiction, For example, the torque pulses common to harmonic drives can be compensated for within the perfect joint. The perfect joint is part of the lowest NASREM level. The processing provided theoretically achieves a "perfect" torque, where the outputted torque matches the commanded torque.

The Individual Joint Controller (IJC) implements a simple control law to allow joint by joint servoing of the manipulator.

The IJC corresponds to NASREM level, and provides a functional redundancy to the higher level telerobot control discussed below. The IJC accepts inputs from a kinematic ally similar mini-master controller. This simplifies the computational requirements on the IJC, by removing the need for coordinate transformations. The IJC does not include any dynamic joint coupling compensation. It basically implements seven parallel, non-interacting control laws, that may be simple PD loops. For this case, roughly 140 operations per cycle are required.

The telerobot controller initially implemented the first three NASREM levels, and could accept commands from a joystick-type element, a mini-master, or higher levels of the model. This level required a computational capability of several MIPS, and an accuracy of 32 bits. Floating point capability was assumed. This controller could perform coordinate transformations in real time, although the computation burden argued for a custom hardware approach to this particular subset of the computations.

The telerobot control system implemented the first 3 (of 7) levels of the NAS-REM model. Further levels could be added later in a phased evolution of the system. For early systems, the human operator provided the functionality of the upper control levels.

The FTS safety system received particular attention, and was specified to be built with class-S parts with flight pedigree, known to be SEU hard, and generally bulletproof.

The approach taken to flight telerobot safety was to use a separate safety watchdog computer system, with a separate low level hardwire control. The separate safety watchdog computer usage did not imply that it is the sole repository of safety responsibility. The implementation of safety was distributed in the system, from the workstation to the robot control computers. At all levels, the system checked the "reasonableness" of actions before they are carried out. The safety computer hardware was implemented in class-S parts that

41

had a flight pedigree, and that had demonstrated a high SEU tolerance. Each safety computer monitored both control computers, both joint controllers, and both manipulators. The safety computers were implemented as a redundant pair, and either could safe the system in a problem situation.

This configuration met the NSTS safety requirements (section 201.Ie (1)) for computer based active processing to prevent a catastrophic hazard. That section stated 'While a computer system is being used to actively process data to operate a payload system with catastrophic potential, the catastrophic hazard must be prevented in a two failure tolerant manner. One of the methods to control the hazard must be independent of the computer system. A computer system shall be considered zero fault tolerant in controlling a hazardous system (i.e., a single failure will cause loss of control), unless the system utilizes independent computers, each executing uniquely developed instruction sequences to provide the remaining two hazard controls." The configuration described satisfied the safety criteria without the use of space qualified parts; however, mission success criteria dictated the use of flight qualified parts for the telerobot control system. For example, an independent series power switching arrangement will satisfy the NSTS safety requirements for three inhibits. However, if either of the relays can fail open, there is an impact on mission success in that the telerobot cannot then be powered. Both the design and the choice of qualified parts were impacted by the criteria of safety and mission success.

The architecture for the flight telerobot was in conformance with JSC's flight directives on flight microprocessor utilization, in that separate computers and flight loads are used for the main controller and the safety system. In this scheme, the safety system can be a fail-safe system, and the main controller can be less than bulletproof. Full two-fault tolerance was provided by the Joint level (manual) analog control, with appropriate switching of joint power. The implementation of the true separation of inhibits and monitoring of inhibits for the switching of control was carefully examined, to ensure that "sneak paths" did not unintentionally remove functional inhibits.

Generally, the communication requirements of the telerobot system, which are bandwidth and maximum latency requirements, could be met by MIL-STD-1553 links. However, certain of the communication links could not, for safety or bandwidth reasons, be shared links. There was a need for certain control and data pathways with a small bounded and deterministic latency on the order of 1-5 milliseconds. A frequency of 200Hz was required for stable force reflection operation with man-in-the-loop. Hardwire links, or more properly, point-point links, could be implemented in optical fiber.

Processor Choice

This section provides a short term view of a practical implementation of the computer architecture at the time. It must be stressed that the architecture derived here is independent of implementation detail. In this section, the requirements were matched to a spectrum of hardware that was real and appropriate for the early 1990's.

The following processors were chosen for this implementation, based on availability, flight experience, and availability of support.

Joint Controller: 80C86 (16-bit)
Safety Computer: 80C85 (8-bit)
Robot Control 80386 (32-bit)
Workstation : 80386 (32-bit)

The workstation and robot control computers were essentially the architecture of the space station standard data processor, SDP-4, then being qualified as part of Work Package 2. The safety computer, although 8 bit, was radiation-hard, SEU hard, and available. We assumed a 5 Mhz clock rate for the 8085, a 3.5 Mhz clock rate for the 8OC86, and a 16 Mhz clock rate for the 80386. Both the 80C85 and the 8OC86 had flown in space. The 80386 and associated math coprocessor 80387 were being qualified for space flight.

The 80C86 used in the Joint controllers needed to be supplemented with an 8087 floating point coprocessor. Although the 80C86 was flight qualified, the 8087 was only available in a mil-spec version.

The Safety computer had to be made from highly reliable, class-S parts and be single event upset free. Since the processing requirements were low, a good candidate was the Harris 80C85 chip. Its LET was greater than 75.

Although parallel processing seems like an obvious thing to do, few such systems have actually been built and few people have had a chance to work with one. There was emerging interest in parallel processing with several new companies beginning to offer commercial systems, but no consensus on the best technical approach and very few actual installations. An accurate assessment of the state of the art is "true parallel processing is highly experimental and still not much use in today's marketplace" (High Technology, Feb. 1987). However, the same article concludes "There are some algorithms in business and nature which are very serial in form, but by and large, if you look at the universe around you, it is mostly parallel".

Microprocessors are inexpensive and proven to be very cost effective when applied individually to problems. The problem comes in paralleling hardware. Even now, most modern microprocessors are not designed to be easily paralleled, and techniques such as shared memory and FIFOs are used. It is easy to transform a compute bound problem into an I/O bound problem by attempting to force a collection of uni-processors into the wrong applications.

In addition, the software tools must support the architecture, in terms of parallel-izing the code, loading the network, and debugging. However, the performance gains of parallel processing are attractive. The architecture scales in both processing power and I/O. But. there is no generic parallel topology: the arrangement of the compute and I/O resources interact heavily with the problem domain.

One promising architecture was that of the Inmos Transputer. This device included both an integer and a floating point unit, memory, and I/O on one chip. I/O was via four bi directional 20 Megabit per second links. In addition, a crossbar switch was available to allow connection of 32 links. The addition of the crossbar switch allowed a variable topology to customize to given problem domains. A flight parallel processor using a transputer array had been proposed. The chip been flow on an ESA experimental payload, and was being considered for onboard data handling of large CCD detector arrays.

In a flight telerobot system, a scalable array of parallel processors with a flexible topology could provide the additional computing horsepower and the I/O bandwidth to implement additional functional levels of the NASREM models for more autonomous operations. Scalable parallel processor arrays could provide the basis for implementation of advanced techniques such as flight expert systems or neural networks.

Hosting an Embedded Flight Expert Systems and Neural Architectures

As an evolutionary approach, the telerobot was to transition from direct hands-on human control to more autonomous operation. This was to take the form of the automation of mundane and repetitive tasks. for example. removing a series of fasteners. by using a record/playback mode. Here. the human operator directs there first operation. and the control system memorizes the steps as a "macro". After that. the human operator guides the manipulator to the next fastener and re-executes the macro. This mode is usually termed supervised autonomy. The supervisor can evolve to be another program.

The expert system used to direct the telerobot in lieu of or in conjunction with a human operator implements the upper levels of the NASREM architecture. It operates in a one to several second "real-time" cycle. and has a 10-30 second planning horizon. An expert system shell facilitates the knowledge capture from the expert human operators. organizes the knowledge (procedural abstraction). directs the operation of the machine. and provides an audit trail of decisions. The telerobot control system becomes then a hybrid of knowledge-based and standard control systems. The control system drives the shell inference engine run time module as a subtask. There is no human interaction with the shell during normal operations. Thus. the requirement is for development tools with a robust interface to other software and the operating environment. An expert system is required that interacts directly with the process data not human input in real time. Several such systems were currently available.

Expert systems generally do not take large amounts of complex processing power, but do require many logical operations on a large database. Expert systems' inference engines can use up large amounts of processing power. The objective of any expert system implementation is to transfer operational and problem solving expertise from a human to a program. Ideally, the program learns in an interactive mode. after an initial download. As has been shown numerous times, cognitive defenses impede knowledge transfer by interviewing. The best knowledge engineers cannot always elicit "why" from a domain expert. because the expert may not be able to verbalize "why" he does a certain action in a given situation. Automation of the knowledge capture process. as well as the knowledge base maintenance is desirable. The expert system section of the control should be able to "look over the shoulder" of the astronaut/operator. and learn to operate equipment by observing. and by example. After a period of learning. it should be able to suggest actions in given situations. and control simple tasks. With experience, it can be given control over increasingly complex tasks.

Several flight expert systems have been proposed and prototyped. In general. the processing power of one EDP seems sufficient as a host. although expert systems do not share a host well with other applications due to their need for large amounts of processing.

Another promising technology in its infancy was neural networks. Instantiated as program simulations on digital computers, neural nets were soon to be directly implemented in analog VLSI. Neural network simulations require floating point calculations. The neural approach, which mimics the mammalian brain process but at a faster rate, provides an inherent parallelism in problem solving. Their major advantage is that they are not programmed. but merely taught by example. They form their own internal representation of the problem space. Thus. they would be a natural adjunct to the telerobot control system. It is important to note that expert systems and neural network techniques are not mutually exclusive. and can be used in cooperative mode. Neural network simulations can also be hosted on EDP-class machines.

The plan for use of the FTS in the assembly and maintenance of Space Station Freedom established early mission requirements for servicing and serviceability.

In the end, the FTS was not built or flown. The Development Test Flight (DTF-1) was canceled. It was a bit too ambitious for the technology of the time, and the funds available. However, the project produced a wealth of information for future projects including the resource requirements estimates, and the definitized space station interfaces, and the servicing requirements imposed on the Station itself proved beneficial for the Astronauts who now had to carry out the plan. At the same time, the Space Station Freedom became the International Space Station, merging with a Russian Project.

Early in the era of space exploration, a series of rover vehicles were sent to the Earth's moon. These were designed as precursors to a manned visit. From the mid-1960's through 1976, there were some 65 unmanned landings on the moon. Now, this is the subject of a private effort, the Google X-prize. The moon is still the subject of intense study, with missions from the United

States, Russia, China, India, the European Union, and Japan. In this section, we focus on rovers of the Lunar surface.

Lunar Rovers

Can we use Teleoperated systems on the Moon with operation from Earth? Yes, but its somewhat of a problem with the light-speed delay. A "person-in-the-loop" quickly gets tired of the delayed response. The best approach is directed autonomy, where the robotic unit is capable of many low-level tasks, with supervision from a distance. If the Lunar Orbital Platform-Gateway (LOP-G) gets implemented, it will be feasible to operate telerobots on the surface from the platform. One major application would be the exploration of lunar lava tubes, for potential use as habitat's and manufacturing and storage centers. Lava tubes could be sealed and pressurized to provide living and work space that would not require a bulky pressure suit, and would be inherently radiation shielded. Once a lava tube is entered, the robotic vehicle must operate autonomously, due to the lack of a communications link. We might also consider radio relays withing the tube, but this gets complex fairly quickly. Similar lava tubes are known exist on Mars.

There were two Soviet Lunar rovers, launched in the period 1969-1977.
.

The 1969 Lunokhod 1A was destroyed during launch, the 1970 Lunokhod 1 and the 1973 Lunokhod 2 landed on the moon, and the 1977 Lunokhod was never launched. The successful missions were in operation concurrently with the Zond and Luna series of flyby, orbiter, and landing missions. The Lunokhods were primarily designed to support the Soviet Manned lunar landings. They automatic remote-controlled robots to explore the surface and return pictures. The moon lander part of the Luna spacecraft for Lunokhods were similar to the ones for sample return missions. The rovers ran during the lunar day, stopping occasionally to recharge batteries via the solar panels. During the 2-week night the rover hibernated until the next sunrise, heated by the radioisotope heater.

Lunokhod 1 was a lunar vehicle formed of a tub-like body with eight independently powered wheels. Its length was 2.3 meters. Lunokhod 1 was equipped with a cone-shaped antenna, a directional antenna, four television cameras, and extendable devices to probe the lunar soil for density measurements and mechanical property tests.

An x-ray spectrometer and telescope, cosmic ray detectors, and a laser were also included. The vehicle was powered by batteries which were recharged during the lunar day by a solar array. During the lunar nights a radioisotope heat source kept the internal components at a survival temperature.

The rover stood 135 cm high and had a mass of 840 kg . It was 170 cm long and 160 cm wide Each wheel had an independent suspension, motor, and brake. The rover had two speeds, 1 km/h and 2 km/h.

Lunokhod 2 was equipped with three slow-scan television cameras, one mounted high on the rover for navigation, which could return high resolution images. These images were used by a five-man team of controllers on Earth who sent driving commands to the rover in real time to operate it in telerobot-ic mode. There were 4 panoramic cameras mounted on the rover.

Interestingly, ownership of Lunokhod 2 and the Luna 21 lander was sold by the Lavoxhkin Association for $68,500. in December 1993 at a Sotheby's auction in New York. The buyer was computer gaming entrepreneur and astronaut's son Richard Garriott. The hardware remains in place on the lunar surface.

Lunabotics Mining Challenge

NASA's Mining Competition was established in 2010, and is ongoing. It challenges college students to apply system engineering principles to mining scenarios, and test the hardware in the Caterpillar Mining Area. Schedule, Budget, and Design Philosophy are parameters for judging. NASA also required K-12 outreach. There are a series of awards in different project areas. Quoting from the announcement, "The Lunabotics Mining Competition is a university level competition designed to engage and retain students in Science, Technology, Engineering and Math (STEM). NASA will directly benefit from the competition by encouraging the development of innovative lunar excavation concepts from universities which may result in clever ideas and solutions that could be applied to an actual lunar excavation device or payload. The challenge is for students to design and build a remote controlled or autonomous excavator (lunabot) that can collect and deposit a minimum of 10 kg of lunar dirt within 15 minutes. The complexities of the challenge include the abrasive characteristics of the lunar surface, the weight and size limitations of the lunabot, and the ability to control the lunabot from a remote control center. Twenty two teams from around the nation are ready to compete at the Kennedy Space Center Astronaut Hall of Fame on May 27-28. These are annual events, with teams selected each year.

"The challenge will be conducted in a head-to-head format, in which the teams will be required to perform a competition attempt using the regolith sandbox and collector provided by NASA. NASA will fill the sandbox with simulated regolith, compact it and place rocks in it. Each competition attempt will occur sequentially. Between each competition attempt, the rocks will be removed, the regolith will be returned to a compacted state and the rocks will be returned to the sandbox. Consideration of prize awards will be based on each team's performance during the official competition attempt. All excavated mass deposited in the collector during the competition attempt will be weighed after completion of the competition attempt. The teams that excavate the first, second and third most lunar regolith mass over the minimum excavation requirement within the time limit will respectively win first, second and third place prizes."

Soviet Luna missions

The Soviet Union launched a series of successful lunar landers, sample return missions, and lunar rovers. The Lunokhod missions, from 1969 through 1977, put a series of remotely controlled vehicles on the lunar surface. Lunokhod-1 was an 8-wheeled rover, operated from Earth. It was the first Rover to land on a body other than Earth. It deployed from the landing platform via a ramp. It was operational for 11 months. The follow-on Lunokhod-2 Rover could transmit live video from the surface, and had a series of soil property instruments. Its tracks were seen by the Lunar Reconnaissance Orbiter in 2010. The Lunokhod-3 rover was built but never launched. It resides at a museum. The first and second rovers remain on the moon, although the second rover was sold in 1993 at a Southby's auction. The buyer was Richard Garriott, son of Astronaut Owen Garriott. As of this writing, he has not picked up his property.

The initial purpose for the Lunokhod series was to scout sites for manned landings, and to serve as beacons. The rover could be used to move one

Cosmonaut at a time on the surface as well. Lunokhod had a group of four television cameras, and mechanical mechanisms to test the lunar soil. There was also an X-ray fluorescence spectrometer, and a cosmic ray detector. The second unit conducted laser ranging experiments from Earth via a corner reflector, and measured local magnetic fields. The rover was driven by a team on Earth in teleoperation mode.

Chinese Yutu Mission

Yutu is the name of the Chinese Lunar Rover, and means *Jade Rabbit*. It was launched in December of 2013. It landed successfully on the moon, but became stationary after the second lunar night. It is a 300 pound vehicle with a selection of science instruments, including an infrared spectrometer, 4 mast-mounted cameras including a video camera, and an alpha particle x-ray spectrometer. The rover is equipped with an arm. It also carries a ground penetrating radar. It is designed to enter hibernation mode during the 2-week lunar night. It does post status updates to the Internet, and still serves as a stationary sensor platform.

Indian Chandrayaan Mission

The Indian/Russian Chandrayaan-2 mission includes an orbiter and a robotic lander, with a proposed launch date in 2018. The design is unique in having been selected from student proposals. The lander will be a 6-wheeled, solar-powered robot rover. A previous mission, in 2008, did a soft landing on the Moon, and confirmed the presence of water ice.

Google Lunar X-Prize

This is a lunar robotics competition, organized by the X-Prize Foundation in 2007, and is valid through 2015. It requires a team to develop and demonstrate a robot on the moon that travels at least 500 meters, and transmits back high definition video. The prize for this is $20 million. If accomplished, this would be the first vehicle to operate on the lunar surface since 1976, and the first non-governmental effort. Another goal is to capture images of Apollo hardware on the moon, verifying the presence of water ice, or surviving through the 2-week long lunar night.

This effort was originally to be funded by NASA, but that would have limited the competition to United States Teams. The X-Prize Foundation, funded by Google, has no such restrictions. More than thirty international teams were officially working on this effort.

No one won this contest.

Cube rover

In the same sense that the standardized Cubesat revolutionized small satellite architecture, the CubeRover is attempting the same for space robots. It is a standardized modular planetary rover, developed by Astrobotic Technology, along with Carnegie Mellon University. The first unit is slated to be on the lunar surface in 2020.

The author challenged students in a summer session to use the Cubesat architecture as the basis of a lunar mission. This was to involve a lander, with multiple mini-rovers. The wheeled rovers would return to their "mothership" to ride out the lunar 2-week night, and explore during the day. The compute and data system architecture of the lander would be Cubesat based. There was also to be a Cubesat in Lunar orbit to relay data, as the mission was baselined for the lunar backside.

The student team designed and built the rover, and paired it with the satellite control center software, Cosmos, from Ball Brothers. It was tested at the Robotics facility of the National Institutes of Standards and Technology, and was found to operate as desired.

Hubble Servicing Missions

There were five servicing missions to the Hubble Space Telescope between 1993-2009. These covered the addition of adaptive optics to correct the main mirror flaw, change-out of some instruments, replacing failed components, and updating the flight computer.

Servicing mission one in 1993 involved 7 astronauts, the Shuttle's telerobotic arm, and hundred specialized tools. The arm, operated from the shuttle's aft deck, was used to capture the spacecraft, and maneuver it onto the servicing platform(Flight Support System) in the Shuttle bay. This was because the bus side of the HST used the MultiMission Modulat Spacecraft (MMS) architecture. Among other things, a co-processor was added to the Rockwell DF-224 computer. The co-processor had dual redundant 80386/80387 processor/numeric processor pairs, each with 1MB of RAM and 256kB EEPROM, plus 384kB of non-alterable permanent ROM. The mission was a success.

Mission two in 1997, used the same procedures, and replaced some instruments and a tape recorder with a new solid state memory unit.

Mission 3A went in December of 1999, and responded to the failure of 3 of the 6 onboard gyros. The set of six were replaced, as well as a fine guidance sensor, and the main computer. The old computer, a DF-224, was replaced by a new unit, some 20 times faster, and with 6 times the memory. It had three rad-hard Intel 486 processors running at 25MHz, each with 2MB of SRAM and 1MB of EEPROM. It is still operating as of this writing.

Mission 3B in 2002, brought a new instrument, and improved cooler for one of the instruments, and a change-out of the solar panels.

The Shuttle Columbia disaster almost spelled the end of further servicing missions. A study in 2004 by GSFC came to the conclusion that a fully robotic servicing mission was not currently feasible. A new NASA administrator remove the ban on STS servicing missions. In the mean time, the Hubble's main data handling unit failed, bringing science to an abrupt stop. Service Mission 4 replaced the faulty unit in 2009, and added two additional instruments. They also installed the Soft Capture and Rendezvous System, which will enable future robotic missions.

SMM Servicing Mission

The Solar Maximum mission took extensive data on the Sun's corona in 1980, until an electronics failure. The SMM was built to the MultiMission Modular Spacecraft (MMS) standard, so it was designed to be service-able in space. It suffered several fuse problems, leading to degraded attitude, using magnotorquers only. At this point only three of the seven observing instruments were usable. It was in standby mode in orbit for three years before becoming the first spacecraft to be serviced in space, April of 1984, with Shuttle mission STS-41-C. After several failed attempts to grab the spacecraft by an astronaut, the damaged satellite was captured by the Shuttle's RMS arm, and placed on the Flight Support System in the Shuttle bay, for repairs. These repairs were successfully completed, adding five years to the satellites working life. The spacecraft reentered the atmosphere and burned in December of 1989, taking some of the author's best flight software with it.

In 2004, the Dextre robot was considered for the SMM repair mission, but NASA proceeded with the crewed Shuttle approach instead.

The most interesting thing happened afterwards, though. NASA was able to do the repair on the SMM engineering model at GSFC, using a commercial grade Puma robot arm, both in tele-operation, and full robotic mode.

Mars Rovers

The Mars Rovers are the rock stars of planetary exploration. Operated by NASA's Jet Propulsion Laboratory, these semi-autonomous units wander the surface of the Red planet, seeking interesting geological sites and signs of life. Operating on the surface of Mars presents some unique challenges: the available solar power is greatly reduced from that available on Earth, the temperatures reach extremes of cold, and there is significant wind-borne dust. The initial goal of the rovers was to traverse 200 kilometers in 400 days. Just getting to the surface of Mars from Earth is difficult. There haes been 6 rover missions t Mars.

The one-way light time between Mars and Earth varies from 8 to 45 minutes. The communication shares the use of the three stations of NASA's Deep Space Network with many other ongoing planetary missions. Sometimes, when the geometry is wrong, no communication is possible. Tele-operation from Earth is not feasible. Directed autonomy works, but is a very limited mode. Thus, the rovers have to carry the computation power they need with them.

The Mars Rovers use vision systems to map their environment, and plan their moves. An odometer keeps track of distance traveled. It is estimated that between 50 and 500 million instructions must be executed to move one meter.

Most onboard power is consumed by this computation burden than by actually moving. The onboard computer available in 1987 managed around 20 mips. There was a trade-off in Rover speed, based on the time to compute the move. The Mars Rovers operate with high-level directives, and do local path planning and obstacle avoidance. Figures of merit are watts per mips, the amount of computation that can be done per unit of power, and watts per kilogram-meter-second, the amount of movement that can be done per unit of power.

There is no GPS system around Mars, and surface rovers need absolute position determination. The magnetic field is low, and a compass sensor is not feasible. Inertial sensor, upgraded by occasional celestial fixes (suing the onboard camera) provide the needed position. The lander can be used as a fixed reference point, as was done for the earlier Sojourner.

Mars Rover Sojourner

Mars presents a harsh environment. Temperature ranges are from −40 °C to +40 °C, and there is a high radiation environment due to Mars' lack of a magnetic field. There is low atmospheric pressure, high winds, and sand storms.

The Mars Pathfinder mission landed on Mars on July 4, 1997. It carried a Rover named Sojourner, which was a 6-wheeled design, with a solar panel for power, but the batteries were not rechargeable. The rest of the lander served as a base station. Communication with the rover was lost in September. The Rover used a single Intel 80C85 8-bit CPU with a 2 MHz clock, 64k of ram, 16 k of PROM, 176k of non-volatile storage, and 512 kbytes of temporary data storage. It communicated with Earth via the base station using a 9600 baud UHF radio modem. The communication loss leading to end of mission was in the base station, while the Rover remained functional. The Rover had three cameras, and an x-ray spectrometer.

The computer in the mission base station on Mars was a single RS-6000 CPU, with 1553 and VME buses. The software was the VxWorks operating system, with application code in the c language. The base station computer experienced a series of resets on the Martian surface, which lead to an interesting remote debugging scenario.

The operating system implemented pre-emptive priority thread (of execution) scheduling. The watchdog timer caught the failure of a task to run to completion, and caused the reset. This was a sequence of tasks not exercised during testing. The problem was debugged from Earth, and a correction uploaded.

The cause was identified as a failure of one task to complete its execution before the other task started. The reaction to this was to reset the computer. This reset reinitialized all of the hardware and software. It also terminates the execution of the current ground commanded activities.

The failure turned out to be a case of priority inversion (how this was discovered and corrected remotely is a fascinating story – see refs.) The higher priority task was blocked by a much lower priority task that was holding a shared resource. The lower priority task had acquired this resource and then been preempted by several medium priority tasks. When the higher priority task was activated, it detected that the lower priority task had not completed its execution. The resource that caused this problem was a mutual exclusion semaphore used to control access to the list of file descriptors that the select() mechanism was to wait on.

The Select mechanism creates a mutex (mutual exclusion mechanism) to protect the "wait list" of file descriptors for certain devices. The vxWorks pipe() mechanism is such a device and the Interprocess Communications Mechanism (IPC) used was based on using pipes. The lower priority task had called Select, which called other tasks that were in the process of setting the mutex semaphore. The lower priority task was preempted and the operation was never completed. Several medium priority tasks ran until the higher priority task was activated. The low priority task attempted to send the newest high priority data via the IPC mechanism which called a write routine. The write routine blocked, taking control of the mutex semaphore. More of the medium priority tasks ran, still not allowing the high priority task to run, until the low priority task was awakened. At that point, the scheduling task determined that the low priority task had not completed its cycle (a hard deadline in the system) and declared the error that initiated the reset. The reset had the effect of wiping out most of the data that could show what was going on. This behavior was not seen during testing. It was successfully debugged and corrected remotely by the JPL team.

References
http://www.nasa.gov/mission_pages/mars-pathfinder/
http://research.microsoft.com/en-us/um/people/mbj/Mars_Pathfinder/

Mars Exploration Rover *Spirit* (MER-B) & *Opportunity* (MER-A)

The MER's were launched in 2003. Opportunity landed successfully at Meridiani Planum in January 2004, three weeks after Spirit had landed on the other side of the planet.

The MER's are six-wheeled, solar-powered robots that stand 1.5 meters high, 2.3 meters wide and 1.6 meters long. They weigh 180 kg..

When fully illuminated, the MER solar arrays generate about 140 watts for up to four hours per Martian day (sol) The rover needs about 100 watts to drive. The power system include two rechargeable lithium-ion batteries weighing 16 pounds each, that provide energy when the sun is not shining. Over time, the batteries degrade and are not be able to recharge to full capacity. Besides dust storms that degrade the solar panels, there was a morning ground fog, discovered by the Viking Landers. This bodes well for finding water ice on Mars.

At night, the rovers are heated by eight radioisotope heater units (RHU), which continuously generate 1 watt of thermal energy each from the decay of the radioisotopes. At night, the temperature can drop to -40 degrees C.

The Rovers can talk to the Deep space Network (DSN) on Earth directly, or via the Mars Odyssey or Mars Global Surveyor spacecraft in Mars orbit.

The MER's had a 90-day primary mission, but they operated considerably longer than that. Dust storms reduced the efficiency of the solar panels. In 2009, after *Spirit* was stuck in soft soil for 9 months, it was left as stationary science platform, and is now in low-power hibernation mode. *Opportunity* was still working after 7 years on the surface. The onboard computer uses a 20 MHz RAD6000 CPU with 128 MB of DRAM, 3 MB of EEPROM, and 256 MB of flash memory. It uses the VxWorks operating system, and includes a 3-axis inertial measurement unit. MER's use an onboard path planner, and supervisory control. They each have an arm, with multiple tools. So far, they have returned over 80,000 images of the surface, and considerable information on the surface characteristics and weather.

The Spirit unit became stuck in 2009, and engineers were unable to free it after 9 months of trying. It was re-tasked as a stationary sensor platform. Contact was lost in 2010. The Opportunity Rover lasted until February of 2019, after some 15 years of service

In June of 2018, a massive Martian dust storm covered Curiosity, most importantly its solar panel, with a large amount of duct. The last communications was on June 10.

Mars aircraft robots have been studied, with a goal of covering more area than rovers. However, the thin atmosphere of Mars means much larger wing sizes and other challenges to flying craft.

The next mission in 2021 will include the 2020 Rover, which has a robotic helicopter. It will be an eye-in-the-sky, looking out for hazards, planning a path, and see things that the rover's camera can't. It will be autonomous in operation. It is a technology demonstration, planned to fly five times, during the early mission. The copter blades are a meter in diameter, and it has two counter-rotating sets. Compasses can't work on Mars due o the low magnetic field, so it will use solar tracking abd inertial guidance. It will have its own solar panels. It is carried under the rover. It is dropped to the ground, and the rover moves some distance away so it can ascend. It run linux.

Using the experience and lessons-learned from previous missions since 1997, NASA plans a more capable rover mission to be launched to Mars for 2020. The overall plan for Mars call for focusing initially on a search for water, both as a possible precursor to and supporter of life, and, at the same time, prepare for future human habitation by understanding the Martian environment.

The mission concept was announced in 2012. The Rover will be a follow-on design to Curiosity, with an upgraded sensor payload. The landing system design from Curiosity is expected to be re-used, as well as spare parts from the earlier mission. A sample return-to Earth is contemplated.

Mars Science Laboratory

The Mars Science Laboratory's lander, named *Curiosity*, landed successfully on the Martian surface on August 6, 2012. It had been launched on November 26, 2011. It's location on Mars is the Gale crater, and was a project of NASA's Jet Propulsion Laboratory. The project cost was around $2.5 billion. It is designed to operate for two Martian years (sols). The mission is primarily to determine if Mars could have supported life in the past, which is linked to the presence of liquid water.

The Rover vehicle weights just about 1 ton (2,000 lbs.) and is 10 feet long. It has autonomous navigation over the surface, and is expected to cover about 12 miles over the life of the mission. The platform uses six wheels The Rover Compute Elements are based on the BAE Systems' RAD-750 CPU, rated at 400 mips. Each computer has 256k of EEprom, 256 Mbytes of DRAM, and 2 Gbytes of flash memory. The power source for the rover is a radioisotope thermal power system providing both electricity and heat. It is rated at 125 electrical watts, and 2,000 thermal watts, at the beginning of the mission. The operating system is WindRiver's VxWorks real-time operating system.

The computers interface with an inertial measurement unit (IMU) to provide navigation updates. The computers also monitor and control the system temperature. All of the instrument control, camera systems, and driving operations are under control of the onboard computers.

Communication with Earth uses a direct X-band link, and a UHF link to a relay spacecraft in Mars orbit. At landing, the one-way communications time to Earth is 13 minutes, 46 seconds. This varies considerably, with the relative positions of Earth and Mars in their orbits around the Sun,. At certain times, when they are on opposite sides of the Sun, communication is impossible.

The science payload includes a series of cameras, including one on a robotic arm, a laser-induced laser spectroscopy instrument, an X-ray spectrometer, and x-ray diffraction/fluorescence instrument, a mass spectrometer, a gas chromotograph, and a laser spectrometer. In addition, the rover hosts a weather station, and radiation detectors. There is cooperation between in-space assets and ground rovers in sighting dust storms by the meterological satellite in Mars orbit.

Mars Science Laboratory Curiosity

The Mars Science Laboratory (MSL) landed successfully on the Martian surface on August 6, 2012. It had been launched on November 26, 2011. It's location on Mars is the Gale crater, and was a project of NASA's Jet Propulsion Laboratory. The project cost was around $2.5 billion. It is designed to operate for two Martian years (sols). The mission is primarily to determine if Mars could have supported life in the past, which is linked to the presence of liquid water. NASA has a Planetary Protection Officer assigned to Mars, to make sure we don't contaminate it with Earth life, and to protect any lifeforms that we may find there. So far, Earth is the only place we are aware of that has life. Is life on Earth unique in the Universe, or it is common? Mars may hold that answer for us.

Gale Crater is a location of interest. It's about 96 miles in diameter, and has a mountain, Aeolis Mons, in the center that is 18,000 feet high.

The Rover vehicle *Curiosity* weights just about 1 ton (2,000 lbs.) and is 10 feet long. It has autonomous navigation over the surface, and is expected to cover about 12 miles over the life of the mission. The platform uses six wheels The Rover Compute Elements are based on the BAE Systems' RAD-750 CPU, rated at 400 mips. Each computer has 256 Mbytes of RAM, and 2 Gbytes of flash memory. The power source for the rover is a radioisotope thermal power system providing both electricity and heat. It is rated at 125 electrical watts, and 2,000 thermal watts, at the beginning of the mission. The operating system is WindRiver's VxWorks real-time operating system. The vehicle was assembled and tested at NASA's Goddard Space Flight Center, and shipped to lead center JPL. The landing location in Gale Crater was named Bradbury Landing, after the science fiction writer, Ray Bradbury. Mars figured heavily in his writings. Gale Crater is named after an Australian amateur astronomer, Walter Gale. There is some evidence that the crater was once filled with water.

The computers interface with an inertial measurement unit (IMU) to provide navigation updates. The computers also monitor and control the system temperature. All of the instrument control, camera systems, and driving operations are under control of the onboard computers.

Communication with Earth uses a direct X-band link, and a UHF link to a relay spacecraft in Mars orbit. At landing, the one-way communications time to Earth was 13 minutes, 46 seconds. This varies considerably, with the relative positions of Earth and Mars in their orbits around the Sun.

The science payload includes a series of cameras, including one on a robotic arm, a laser-induced laser spectroscopy instrument, an X-ray spectrometer, and x-ray diffraction/fluorescence instrument, a mass spectrometer, a gas chromotograph, and a laser spectrometer. In addition, the rover hosts a weather station, and radiation detectors. There is cooperation between in-space assets and ground rovers in sighting dust storms by the meteorological satellite in Mars orbit.

In 2013, NASA uploaded a software upgrade to Curiosity's operating System. Overall, it took a week to install.

Curiosity's exploration of the ancient lake bed, known as Gale Crater resulted in some new discoveries that NASA released on June 7, 2018. It found organic molecules, particularly methane, below the surface. Curiosity has a sampling drill (that, unfortunately, is limited to 5 cm.), a mass spectrometer, and a gas chromograph. On Earth, most methane is from biological processes. It can be produced by in-organic processes, however. Scientists have also discovered a season pattern in the amount of methane in the atmosphere, in the amount of a factor of three, that may point to sub-surface storage.

As this book went out to be published, the rover is still operational on the Martian surface.

The source of the Mars methane has to be resolved. This will be a major goal of NASA's and ESA's next landers. A large underground reservoir of methane could be very useful for return-trip rocket fuel.

The European Space Agency, working with the Russians, conducted a 2016 launch of a project called ExoMars in a search for evidence of the existence of past life. Schiaparelli, a stationary lander will use a companion spacecraft in Mars orbit to communicate with Earth. It includes a panoramic camera, a drill assembly, and an analytical laboratory to detect the pretense of life. This lab includes MOMA (Mars Organic Molecule Analyzer), and infrared imaging spectrometer, and a Raman spectrometer. The rover will also have a ground penetrating radar. The Russians are contributing a neutron spectrometer. and a Fourier spectrometer.

Reference: http://smsc.cnes.fr/EXOMARS/

Zero Robotics Competition

This program involves a series of robots already on the International Space Station called SPHERES (Synchronized Position Hold, Engage, Reorient Experimental Satellites). These have a mass of around ten pounds, and a diameter of 8 inches. They use twelve CO_2 thrusters for movement, and are battery powered. They were developed at the MIT Space systems Laboratory as a testbed for control, autonomy, and metrology for distributed spacecraft and docking missions. The SPHERES were inspired by the Training Remotes from the Star Wars films. There are three SPHERES, in different colors.

As a team, they can control their relative their relative position and orientation. They had been tested aboard KC-135 aircraft flying zero-gravity flight paths, and were delivered to the International Space Station (ISS) in 2006.

The NASA/MIT Competition allows teams to develop software for the SPHERES, and test it in a simulation environment. Selected teams test their software on SPHERES on an air-bearing floor facility. In December 2011, a few teams tesedt their code and algorithms on the unit onboard the ISS.

The Shuttle's ARM

The manipulator arm on the Space Shuttles and the International Space Station were developed in Canada by SPAR Aerospace under a 1980 agreement. These are teleoperation robots.

Called the Remote Manipulator System (RMS), each Shuttle carry's one, and the Space Station had a modern variation. The RMS is teleoperated from the Aft flight deck of the Shuttle. It is normally stowed along the sill of the cargo bay, on the shuttle's left side. It is 50 feet long, and can handle a mass of 60,000 pounds. It has been demonstrated to be safe when used in conjunction and proximity to humans (astronauts) also doing servicing tasks. It is designed to be jettison-able if control fails on-orbit, so the cargo bay doors of the Shuttle can be closed for re-entry.

The RMS had a 2-axis shoulder joint where it attaches to the Shuttle, a single axis elbow, and a 3-axis wrist. The end of the RMS is called the end effector, and is designed to mate with a specific fixture on spacecraft or ORU's.

Designed for zero-G operation, the RMS cannot support its own weight on Earth. An air-bearing floor is used for 2-axis simulations. There were full-scale hydraulic analogs at Johnson Space Center and Goddard Space flight Center, but these could only manipulate full size but inflatable payloads. RMS operations are planned with simulation software.

Robotics on the ISS

This section discusses the use of robotic and telerobotic systems on the International Space Station. On-orbit construction of thhe ISSbegan in 1998, and was completed with a last Shuttle mission in 2011. It is the largest artificial satellite in Earth orbit, and can be seen from the ground with the naked eye. The ISS is a synthesis of several space station modules from the U. S., the Soviets/Russians, the Europeans, and the Japanese. It serves as a laboratory, observatory, and factory in Earth orbit, and is continuously crewed. It has several telerobot arms on he outside, to aid in construction and maintenance. The station currently has multiple robot arms as well as a mobile crane.

AERCam Sprint

The Autonomous Extravehicular Activity Robotic Camera Sprint allows for autonomous inspections of the exterior of the ISS. It is ball-shaped, 14 inches in diameter, and was was first tested on STS-87. It includes cameras, and a thruster system.

Mobile Servicing System

The Mobile Servicing System (MSS) on the International Space Station was produced by MDA Space Missions as a contribution from the Canadian Space Agency to the International mission. It arrived at the Station in 2001. It can move equipment and EVA astronauts around the outside of the station. The system consists of a mobile base that moves on rails outside of the station, on the Integrated Truss Structure. The base is called the Mobile Transporter Cart. The cart transports the Canadarm assembly, which may be itself connected to Dextre, or an EVA astronaut.

The Mobile Servicing System consists of Candarm2, the Mobile Remote Servicer Base, and the Special Purpose Dexterous Manipulator. With the Shuttle-derived latching end effector, the system can latch onto and move large assemblies out the station.

Dextre

The Special Purpose Dexterous Manipulator (SPDM), called Dextre is manufactured by the Canadian firm MacDonald Dettwiler. It is used in conjunction with the Canadarm assembly, and is sometimes referred to as the Canada Hand. It includes 2 short arms and various end-effectors for servicing operations.

It arrived at the Station in May of 2008, on shuttle Mission STS-123. It was assembled in orbit by EVA astronauts from the Station. It can be operated by the onboard crew in telerobotic fashion, or from the ground.
Dextre's arms are 11 feet long, fitted to a body assembly that can bend at the waist. The Canada arm assembly can grasp the Dextre body by means of a standard grapple fixture attached to its back. The ARM can be moved to various worksites around the Space Station. Each of Dextre's arms have an ORU/Tool Change-out Mechanism, that allows various tools to be attached. These include grasping jaws, a socket drive, a camera, light assemblies, and a connector to provide power and a data connection. The robot body includes lights and dual color cameras, as well as a platform to hold ORU's and tools. The actual hand that attached to the Dextre arm is called SARAH, the Self Adaptive Robotic Auxiliary Hand. Dextre gets electrical power from the Canadarm2.

Robonaut

The Robonaut is a circa-1997 NASA Johnson Space Center Dexterous Robotics Laboratory Project to define an Astronaut-equivalent humanoid telerobot for use inside or outside the Space Station. The focus for Robonaut is in dexterity, and safety in working with Astronauts. Being human-form, it can use tools developed for astronauts.

The initial Robonaut design, circa 1996, was to be used as an end-effector on the Station's robotic arm, so it could accomplish EVA tasks. Two versions were built, but none were flown.

The new version, Robonaut–2, was launched to the International Space Station onboard Space Shuttle flight STS-133 in February 2011. His legs were sent up on a later resupply flight, and were attached in 2014. Try that with an Astronaut. The Robonaut became a permanent resident on the station.

Robonaut-2 is a joint NASA-General Motors Project, and represents the first humanoid robot in space.

As an Astronaut-equivalent, the robot will have roughly the same size, strength, and dexterity as an Astronaut. It will use the same tools, handholds, hatches, and such. It has 5-fingered hands with 12 degrees of freedom, which are gloved-hand equivalent.

One or more Robonauts can perform co-operative tasks with astronauts. This aspect has been tested extensively on Earth at JSC. There are over 350 sensors in the Robonaut, and 38 PowerPC computers. The Robonaut is designed to be connected to a station laptop. It currently must be plugged into a station power outlet, but a battery pack for the unit is in development.

The unit weighs 330 pounds on Earth, and has a mainly aluminum structure. There are a total of 42 degrees of freedom in the unit, including 3 in the neck, 7 in the upper arm and wrist, and 12 in the hands. It also has waist rotation. Joints are controlled by servo motors. The fingers have tactile sensing, and integrated load cells in the finger joints. The finger gripping surfaces are a high friction material

Robonaut 's may find additional off-planet work as explorers - keep an eye on this technology. On Earth, a large number of technologies from Robonaut are available to use under license. http://Robonaut.jsc.nasa.gov

Canadarm

Formally called the Remote Manipulator System (RMS), each Shuttle carry's one, and the Space Station has a updated version. These were made in Canada.

Candarm2

The derivative of the Shuttle's Arm, the Remote Manipulator System (RMS) is the Canadarm2 for the Space Station. It was launched in April of 2001. It is 58 feet long, and has seven degrees of freedom. It is capable of manipulating payloads of up to 256,000 lbs on orbit (where inertia, not weight, is the issue). The Arm interfaces with Power Data Grapple Fixtures (PDGF) located on the station. It can move from one PDGF to another , as each end of the arm assembly can be connected. It can also be moved on the Space Stations's railroad, the Mobile Base System, which rides on rails. Inside the Station, crew members at one of three Robotic Work Stations (RWS) located throughout the station, can view and operate the arm. It is operated by dual hand controllers, one for translation, and one for rotation.

The Japanese Experiment Module (JEM, named Kibo) is the largest on the station. It came up on three Shuttle missions. There are six major elements, including the pressurized lab,and the exposed facility, and there is a robot arm.

Robotic Refueling Mission

The Robotic Refueling Mission (RRM) is a joint NASA/Canadian Space Agency Project to test hardware and techniques for refueling spacecraft in orbit. This will include spacecraft that were not specifically designed to be serviced, or refueled. At this point, every satellite in orbit has not specifically been designed for on-orbit service.

The on-orbit demonstration will be done at the International Space Station. An RRM module weighing 250 kilograms would be mounted outside the station habitable area. It will contain a fluid transfer experiment, using some 1.7 liters of ethanol. Inside the module will be four purpose built tools for testing. These include a wire cutter and (thermal) blanket manipulation tool, a safety cap removal tool, a multi-function tool, and a nozzle tool.

The RPM package was delivered to the Station on Shuttle Mission STS-135, the last mission. It was removed from the shuttle cargo bay by 2 astronauts, and placed on a temporary platform. Later, the Space Stations arm assembly moved it to the Express Logistics Carrier-4.

Personnel at the Johnson Space Center in Houston operate the Station's Dextre Telerobot, which consists of two dexterous arms. They use the tools from the RRM module, which latch onto the end of the arms. Additional tools and task boards will be sent to the station later.

In 2012, the RRM had shown that remotely controlled telerobots could perform precise servicing tasks in space, in low-clearance working spaces. A fluid transfer was accomplished in January 2013. All of these operations had been extensively tested and verified on the ground.

The RPM tasks are:

Launch Lock Removal and Vision - The Dextre robot releases the "launch locks" on the four RRM servicing tools. These locks kept the tools secure within the RRM module during the shuttle Atlantis' flight to the International Space Station. Then Dextre's cameras image the hardware in both sunlight and darkness, providing data to develop machine vision algorithms that work against harsh on-orbit lighting.

Gas Fittings Removal - Marking the first use of RRM tools on orbit, Dextre uses the tools to remove the fittings that many spacecraft have for the filling of special coolant gases.

Refueling - After snipping lock wires and removing caps, Dextre is able to access a fuel valve similar to those commonly used on satellites today and transfer liquid ethanol through a sophisticated robotic fueling hose, completing a first-of-its-kind robotic refueling event.

SMA (Sub-miniature A) Cap Removal - Dextre removes the coaxial radio frequency (RF) connector caps that terminate and protect the RF connector while the satellite is in orbit. Access to these connectors would allow a robotic servicer to plug into the data systems of a satellite and better diagnose an internal issue.

Screw Removal - Dextre will acrobatically unscrew satellite bolts (fasteners). RRM draws from its experience with the Hubble Space Telescope servicing mission in its use of a small cage to guide the tool tip and ensure that no fasteners float away.

Thermal Blanket Manipulation - Dextre slices off thermal blanket tape and folds back a thermal blanket to access the contents underneath.

Reference: http://ssco.gsfc.nasa.gov/rrm_tasks.html

Afterword

It is clear that robots and telerobotic systems are taking a lead role in space exploration and operations. They might be the pioneers and the blue (metal) collar workers for us, but people have to get to the other planets of our solar system, and to the stars.

The early work with manipulators such as the Shuttles RMS (arm) and servicing in orbit shows this approach is feasible, as many expensive missions have been recovered. It is time to develop the infrastructure to allow robotic servicing in other orbits, and further out.

In a sense, all satellites are robots – remote explorers. A follow-on book will discuss the topic of autonomous robot repair and servicing ot satellites in Earth orbit. This is an important topic, now that we don't have the Shuttle.

Glossary

Actuator – device which converts a control signal to a mechanical action.

A/D, ADC – analog to digital converter.

ALU – arithmetic logic unit.

Analog – concerned with continuous values acuitychange continuously

AR&D – Autonomous Rendezvous and Docking.

ASIN – Amazon Standard Inventory Number

Async – asynchronous; 2 processes not sharing the same clock.

BAA – Broad Agency Announcement (U. S. Government)

Bus – data channel, communication pathway for data transfer.

CAN – controller area network.

Chip – integrated circuit component.

Clock – periodic timing signal to control and synchronize operations.

CPU – central processing unit.

CRADA – Cooperative Research and Development Agreement (U. S. Government and industry)

CSA – Canadian Space Agency.

Device driver – specific software to interface a peripheral to the operating system.

Dextre - Dexterous Manipulator robot arm, Canadian, on Space Station.

Dof – degree of freedom.

Droid – robot.

ELV – Expendable Launch Vehicle.

EMF – electro-magnetic field or force; voltage.

ESA – European Space Agency

EVA – Extra Vehicular Activity- involving an Astronaut with suit and maneuvering unit in space.

FAR – (US) Federal Acquisition Regulations.

FPGA – field programmable gate array.

FPP – Firm Fixed Price (Contract)

FSS – Flight Support System, structure in Space Shuttle bay to hold spacecraft.

FTS – Flight Telerobotic Servicer.

GEO – geosynchronous Earth orbit.

Giga - 10^9 or $2^{30.}$

GHz – giga ($10^{9)}$ hertz.

GOES – NASA/NOAA Geostationary Operational Environmental Satellite

GNFIR - GSFC Natural Feature Image Recognition System

GPS – global positioning system (U.S.) system of navigation satellites.

Gray - unit of radiation, =100 rad

GSFC – Goddard Space Flight Center, Greenbelt, Maryland. NASA Center for unmanned spacecraft near Earth.

Hz – Hertz, or cycles per second.

IEEE – Institute of Electrical and Electronic Engineers. Professional organization and standards body.

Intelsat - International Telecommunications Satellite Organization.

Interrupt – an asynchronous event to signal a need for attention (example: the phone rings).

IP – Intellectual Property.

IR – infrared, 1-400 terahertz. Perceived as heat.

ISBN – International Standard Book Number.

ISRS – In-space robotic servicing.

ISS – International Space Station.

LEO – low Earth orbit.

LV – launch vehicle

MMS – MultiMission Modular Spacecraft.

MMU – manned maneuvering unit – for EVA astronauts.

NASREM - NASA/NBS Standard Reference Model for Telerobot Control System

NASA – National Aeronautics and Space Administration (USA)

NBS - National Bureau of Standards, now NIST.

NIST – National Institutes of Standards and Technology.

NOAA – National Oceanographic and Atmospheric Administration. (USA)

NSSC-1 NASA Standard Spacecraft Computer-1.

Open source – methodology for hardware or software development with free distribution and access.

ORU – Orbital Replacment Unit.

PDGF – Power Data Grapple Fixture, on the Space Station.

RCS – robot control system.

RFI – Request for Information

RMS – remote manipulator system – the Shuttles "arm"

RRM – Robotic Refueling Mission.

RSV – RESTORE Servicing Vehicle.

RTOS – real-time operating system.

RWS – Robotic Work Station, on Space Station.

SARAH – Self Adaptive Robotic Auxiliary Hand, (on Dextre).

Sensor – a device that converts a physical observable quantity or event to a signal.

SCADA – Supervisory Control and Data Acquisition – for industrial control systems.

Serial – bit by bit.

SM – servicing mission.

SMA – sub-miniature type A connector.

SMM – Solar Maximum Mission, an MMS mission.

SPDM – Spercial Purpose Dexterous Manipulator on Space Station, aka Dextre

SSCO – Satellite Servicing Capabilities Office, NASA, GSFC.

STS – Space Transportation System (USA) Shuttle.

System – a collection of interacting elements and relationships with a specific behavior.

System of Systems – a complex collection of systems with pooled resources.

TDRSS – Tracking and Data Relay Satellite.

Telerobot – a robotic system with a human in the loop.

Transceiver – receiver and transmitter in one box.

Watchdog – hardware/software function to sanity check the hardware, software, and process; applies corrective action if a fault is detected; fail-safe mechanism.

Bibliography

Ambrose, R. O.; Aldridge, H.; Askew, R. S.; Burridge, R.; Bluethman, W.; Diftler, M. A. ; Lovchik, C.; Magruder, D.; Rehnmark, F. "Robonaut: NASA's Space Humanoid," IEEE Intelligent Systems Journal, Vol. 15, No. 4, pp. 57–63, July/Aug. 2000.

Arkin, Ronald C. "Governing Lethal Behavior: Embedding Ethics in a Hybrid Deliberative/Reactive Robot Architecture," Technical Report GIT-GVU-07-11, Georgia Institute of Technology.

Austin, Edmund; Fong, Chung P. "Teleoperated Position Control of a Puma Robot," 1987, JPL, avail: ntrs.nasa.gov

Aziz, S. "Lessons learned for the STA-120/ISS 10A robotics operation", Acta Astronautica, ISSN-00945765, Vol. 66, pp 157-165.

Backes, P. G. Long, M. K. Steele, R.D. "The Modular Telerobot Task Execution System for Space Telerobotics," IEEE International conference on Robotics and Automation, 1993 Proceedings.

Bräunl, Thomas; *Embedded Robotics: Mobile Robot Design and Applications with Embedded Systems* Springer; 2nd ed. edition (July 28, 2006) ISBN-3540343180.

Clancey, William J. *Working on Mars: Voyages of Scientific Discovery with the Mars Exploration Rovers* (MIT Press), 2012, ISBN-026201775X.

Coleshill, E., et al "Dextre: Improving maintenance operations on the International Space Station," Acta Astronautica 64, pp 869-874

Cress, John D.; Mantooth, H. Alan; *Extreme Environment Electronics*, 2012, CRC Press, 1st ed, ISBN-1439874301.

Diftler; A.; Culbert, C. J.; and Ambrose, R. O. ; "Evolution of the NASA/DARPA Robonaut Control System," in *IEEE International Conf. Robotics Automation*, pp. 2543–2548, 2003.

Dudek, Gregory Jenkin, *Computational Principles of Mobile Robotics*, 2000, Cambridge University Press, ISBN 0521568765.

Dvorak, Dan (task lead), *NASA Study, Flight Software Complexity,* Sept. 2008, CL#08-3913.

Ellery, Alex; *An Introduction to Space Robotics*, 2000, Springer, ISBN 9781852331641.

Erickson, Jon D. *Manned Spacecraft Automation and Robotics*, Artificial Intelligence and Information Sciences Office, NASA Lyndon B. Johnson Space Center, Proceedings of the IEEE Vol. 75, No. 3, March 1987, pp. 417-426 .

Everett, H. R. *Sensors for Mobile Robots Theory and Applications,* 1995 A. K. Peters Ltd. ISBN 1-56881-048-2.

Everett, Hobart R. "A Microprocessor Controlled Autonomous Sentry Robot," October 1982, Thesis, Navel Postgraduate School, Monterey, CA, A125239.

Fortescue, Peter and Stark, John *Spacecraft System Engineering*, 2nd ed, Wiley, 1995, ISBN 0-471-95220-6.

Gennery, Donald B., Litwin, Todd, Wilcox, Brian, Bon, Bruce; "Sensing and Perception Research for Space Telerobotics at JPL, Proc. IEEE International Conference on Robotics and automation, Raleigh, NC, March 31-April 3, 1987.

Goldberg, S., Maimone, M., L. Matthies, L. ;"Stereo Vision and Rover Navigation Software for Planetary Navigation, IEEE Aerospace Conference, March 2002, pp. 2025-2036, Big Sky, Montana.

Hirzinger, G. "Robots in Space - A Survey," Advanced Robotics, Vol 9, n6, 2994, pp 625-651.

Holmes-Siedle, A. G. and Adams, L. *Handbook of Radiation Effects*, 2002, Oxford University Press, ISBN 0-19-850733-X.

Kader, Jac B and Loftin, R. Bowen *Standards for Space Automation and Robotics*, AIAA 92-1515, AIAA Space Programs and Technologies conference, March 24-27, 1992, Huntsville, AL.

Kim, W. S. "Graphical Operator Interface for Space Telerobotics," 1993 IEEE conference of Robotics and Automation. ISBN 0-8186-3450-2.

Korf, Richard E. "Space Robotics", Carnegie-Mellon University, Robotics Institute, CMU-RI-TR-82-10, ASIN: B0006Y2KJE.

Landis, G. "Teleoperation from Mars Orbit: A Proposal for Human Exploration," *Acta Astronautica, Vol. 61,* No. 1, 59-65 (Jan. 2008); also paper IAC-04-IAA.3.7.2.05, 55th International Astronautical Federation Congress (2004).

Landis, Geoffrey A. "Robots and Humans: Synergy in Planetary Exploration," NASA John H. Glenn Research Center, ACTA Aeronautica, V. 55, Issue 12, December 2004, pp 985-990.

Lane, J. Corde, Carignan, Craig R. and Akin, David L. "Advanced Operator Interface Design for Complex Space Telerobots, Autonomous Robots, V. 11, n1, pp 49-58.

Lauderbaugh, "Critical Issues in Robot-Human Operations During the Early Phases of the Space Station Program," C-STAR Final Report, 26 Feb 1988, NAG5-939.

Launius, Roger D. and McCurdy, Howard E.; *Robots in Space: Technology, Evolution, and Interplanetary Travel* (New Series in NASA History), The Johns Hopkins University Press; January 7, 2008, ISBN-0801887089.

Levenson, Nancy G. *Safeware, System Safety and Computers*, 1995, Addison Wesley, ISBN 0-201-11972-2

Liang, Bin, Li, Cheng, Xue, Lijun, Qiang, Wenyi; "A Chinese Small Intelligent Space Robotic System for On-Orbit Servicing", 2006: Intelligent Robots and Systems, IEEE ISBN 1-4244-0258-1.

Messenger, G. C. *The Effects of Radiation on Electronic Systems*, 2014, Springer, ISBN- 9401753571.

Montgomery, Raymond C. *Space Robotics - Recent Accomplishments and Opportunities for Future Research*, 1992, NASA-TM-107675.

Morrison, Jack and Nguyen, Tam; On-Board Software for the Mars Pathfinder Microrover, Second International Conference on Low-cost Planetary Missions, Laurel, MD, 1996

Pons, Jose L. *Wearable Robots: Biomechatronic Exoskeletons*, Wiley, 2008, ISBN 9780470512944.

Reed, Benjamin *Overview of NASA's In Space Robotic Servicing*, 2015, ASIN-B01BB1EDJK.

Rokey, M. Grenander, S. "Planning for Space Telerobotics: the Remote Mission Specialist," June 1990, IEEE Expert, Vol 5, issue 3 pp 8-15.

Sagan, Carl *Machine Intelligence and Robotics: Report of the NASA Study Group*, June 1977-Sept. 1979, NASA TM-82329, N81-21769. NTIS Control Number 120251233.

Schenker, P. S. "NASA Research and Development for Space Telerobotics," IEEE Transitions on Aerospace and Electronic Systems, Sept 1988, v 24 n 5 pp 523-534.

Schroer, Bernard J. "Telerobotics Issues in Space Application, "Robotics and Autonomous systems, V. 4 Issue 3, Nov 1988, pp 233-344.

Skaar, Steven B. *Teleoperation and Robotics in Space*, 1994, AIAA, ISBN 1563470950.

Squyres, Steve *Roving Mars Spirit, Opportunity and the Exploration of the Red Planet,* 2005 Hyperion Books, ISBN 1-4013-0149-5.

Stakem, Patrick H. "Advanced Computational Architecture for Flight Telerobotic Servicers", Satellite Services Workshop IV, June 21- 23, 1988, Johnson Space Center, Texas.

Stakem, Patrick H. *History of Spacecraft Computers from the V-2 to the Space Station*, 2011, PRB Publishing, ASIN B004L626U6.

Stakem, Patrick H. "The Applications of Computers and Microprocessors Onboard Spacecraft, NASA/GSFC 1980.

Stephens, Dr. K. Dean *Proxy Robotics: To the Moon and Beyond*, 2013, ASIN-B00B8CL08U.

Stoney, William E. *Cooperative Intelligent Robotics in Space II*: 12-14 November 1991, Boston, MA, SPIE, ISBN 0819407496.

Tomayko, James "Computers in Spaceflight, The NASA Experience," 1987. http://history.nasa.gov/computers/contents.html

Truszkowski, Walt *Autonomous and Autonomic Systems: With Applications to NASA Intelligent Spacecraft Operations and Exploration Systems,* Springer; 1st Edition. edition, 2009, ISBN-1846282322.

Truszkowski, Walt; Clark, P. E.;, Curtis, S.; Rilee, M.   Marr, G. *ANTS: Exploring the Solar System with an Autonomous Nanotechnology Swarm.* J. Lunar and Planetary Science XXXIII (2002).

Vertesi, Janet *Seeing Like a Rover: How Robots, Teams, and Images Craft Knowledge of Mars,* 2015, ISBN-022615596X.

Weisbin, C. R. and Montemerlo, M. D., "NASA's Telerobotics Research Program," Applied Intelligence, V. 2, n 2, pp 113-125.

Wiens, Roger *Red Rover: Inside the Story of Robotic Space Exploration, from Genesis to the Mars Rover Curiosity,* 2013, ISBN-0465055982.

Wilcox, Brian H.; "Robotic Vehicles for Planetary Navigation," Journal of Applied Intelligence, 2nd Qtr 1992, Kluwer Academic Publishers, Boston MA., 1992, pp. 181-193.

Wilcox, Brian H. and Gennery, Donald B.; "Mars Rovers for the 1990's," J. British Interplanetary Society, Vol. 40 pp 484-488, 1987.

Williams, Robert L. *Automation and Robotics for Space-Based Systems*, 1991: Proceedings of a Workshop, NASA-Langley Research Center.

Wilson, Kevin T. "Analysis of 32-bit Processors for Space system Applications," Feb. 2, 1994, innovative systems & technologies corporation (sic).

Xu, Yangshen; *Space Robotics: Dynamics and Control*, Kluwer, 1993, ISBN 0792392663.

Young, Anthony *Lunar and Planetary Rovers: The Wheels of Apollo and the Quest for Mars*,2007, ISBN- 0387307745.

Resources

Proceedings of 1987 Goddard Conference on Space Applications of Artificial Intelligence (AI) and Robotics, May 13-14, 1987. NASA Goddard Space Flight Center, Greenbelt, Maryland.

The Future of Telerobotics, *Robotics World*, Summer 1996.

WTEC Panel Report on International Assessment of Research and Development in Robotics, George Bekey, Robert Ambrose, Vijay Kumar, Art Sanderson, Brian Wilcox, Yuan Zhen, January 2006, Baltimore, MD.

Space Robotics in Japan, January 1991, Japanese Technology Evaluation Center (JTEC), Loyola College in Maryland.

Design for On-Orbit Spacecraft Servicing, 1991, AIAA Guide G-042, 1991.

"The Mars Rover," Unmanned Systems, Published by the Association for Unmanned Vehicle Systems, Summer 1986.

An Independent Study of Automation & Robotics for the National Space Program, 1985, NASA.

Space Robotics (SPRO '98): A Proceedings Volume from the IFAC Workshop, St. Hubert, Quebec, Canada, 19-22 October 1998. 1999, Pergamon, ISBN 0080430503.

Standard Vocabulary for Space Automation and Robotics, AIAA, 1995, ISBN 1563471329.

Advancing Automation and Robotics Technology for the Space Station Freedom and for the U. S. Economy, ATAC, NASA. NASA-TM-104024.

Selected Topics in Robotics for Space Exploration: Proceedings of a Workshop sponsored by the NASA Langley Research Center and the Center for Intelligent Robotic Systems for Space Exploration, March 17-18, 1993. NASA-CP-10131.

"Landing a Humanoid Robot on the Moon in 1000 Days," Project M, Feb 10, 2010. (see wikipedia)

Advanced Automation for Space Missions (1980) Proceedings of the 1980 NASA/ASEE Summer Study, Sponsored by the National Aeronautics and Space Administration and the American Society for Engineering Education.

"Robots for Space: Two Projects", Robotics World, June 1991 v 9 n 2 p 44.

NASA/Goddard Space Flight Center, "On-Orbit Satellite Servicing Study," Project Report, Oct. 2010.
avail: https://sspd.gsfc.nasa.gov/images/nasa_satellite
%20servicing_project_report_0511.pdf

"Critical Issues in Robot-Human Operations During the Early Phases of the Space Station Program", C-STAR, Final Report, 26 Feb. 1988.

In-space robotic servicing https://gameon.nasa.gov/projects-2/archived-projects-2/robotic-satellite-servicing/

wikipedia, various.

NASREM References

Courtesy, NIST

Albus, J.S. Lumia, R. McCain, H. "Hierarchical Control of Intelligent Machines Applied to Space Station Telerobots, IEEE Transactions on Aerospace and Electronic Systems, Sept 1988, V 24 n 5 pp 535-541.

Albus, James S., McCain, Harry G., Lumia, Ron, "NASA/NBS Standard Reference Model for Telerobot Control System Architecture (NASREM)," NIST Technical Note 1235, 1989 Ed.

Albus, James, et al, 4D/RCS: A Reference Model Architecture for Unmanned Vehicle systems, Version 2, NISTIR 6910, NIST, August 2002, at:http://www.isd.mel.nist.gov/documents/albus/4DRCS_ver2.pdf.

Fiala, John "Manipulator Servo Level Task Decomposition," ICG-#002, NIST, Dec. 3, 1987.

Fiala, John "Interfaces to Teleoperation Devices," ICG-#004, Dec. 3, 1987. also, Oct. 1988, NIST Technical Note 1254.

Fiala, John "Notes on NASREM Implementation, December, 1989, NISTIR 89-4215, NIST.

Gazi, Veysel, Moore, Mathew L., Passino, Kevin M., Shakleford, William P., Proctor, Frederick M., Albus, James S. *The RCS handbook: Tools for Real Time Control Systems Software Development,* 2001, Wiley Interscience ISBN 0471435651.

Kent, Ernest W. and Albus, James S. "Servoed world models as interfaces between robot control systems and sensory data," Robotica (1984) v2, pp17-25.

Madhavan, Raj; Messina, Elena R.; Albus, James S. (Co-editors) *Intelligent Vehicle Systems: A 4D/RCS Approach*, Nova Science Publishers (January 15, 2007) ISBN 1600212603.

Stakem, Pat, Lumia, Ron, Smith, Dave, "A Computer and Communications Architecture for the Flight Telerobotic Servicer," June 24, 1988, ICG-#20, Intelligent Controls Group, Robot Systems Division, National Bureau of Standards.

Tarnoff, Nicholas Jacoff, Adam Lumia, Ronald, "World Model Registration for Effective Off-Line Programming of Robots," NIST, Jan. 1990 (3rd. International Symposium on Robotics and Manufacturing (ISRAM) Vancouver, B.C. Canada July 18-20, 1990.

Wavering, Albert J. "Manipulator Primitive Level Task Decomposition," ICG-#003, Jan. 5, 1988.

"Servo Level Control for Manipulation in the NASREM Architecture," Real Time Control Group, Robot Systems Division, NBS (NIST), June 8, 1987.

NIST Bibliography

V.D. Hunt, *Smart Robots: A Handbook of Intelligent Robotic Systems*, Chapman and Hall, 1985, ISBN-041200531X .

E. Johnsen, "Telesensors, Teleoperators, and Telecontrols for Remote Operation," IEEE Trans. on Nuclear Science, Vol. NSI3, 1966, p. 14.

R. Goertz, "Manipulation Systems Development at ANL," Proc. of 12th RSTD Conf., 1964, p.117.

J. Vertut and P. Coiffet, *Teleoperation and Robotics Evolution and Development*, Prentice Hall, 1984, ISBN-9400789130.

C. Flatau, "Compact Servo Master-Slave Manipulation with Optimized Communication Links," Proc. of 17th RSTD Conference 1969, p. 154.

R. Mosher, B. Wendel, "Force Reflecting Electro-Hydraulic Servo Manipulation," Electro Technology, Vol. 66, 1960, p. 138.

J. Charles, J. Vertut, "Cable Controller Deep Submergence Teleoperation System," Proc. of 2nd RMS Conference, 1975.

P. Mosher, "Exploring the Potential of a Quadruped", SAE, SAE Report 690191, 1969.

R. McGhee, "Control of Legged Locomotion Systems," Proc. of 4th Joint Automatic Control Conference, 1977, p. 205.

D.E. Whitney, "Resolved Motion Rate Control of Manipulators and Human Prostheses," IEEE Trans. Man-Machine Systems MMS- 10, 1969, p. 47.

D.E. Whitney, "The Mathematics of Coordinated Control of Prostheses and Manipulators," Journal of Dynamic Systems, Measurement. Control, Dec. 1972, p. 303.

R.P. Paul, "Manipulator Path Control," IEEE Int. Conf. on Cybernetics and Society, New York,

R.P. Paul, "Manipulator Cartesian Path Control," IEEE Trans. Systems, Man, Cybernetics SMC-9, 1979,

R.P. Paul, Robot Manipulators: Mathematics, Programming, and Control, MIT Press, 1981.

R. H. Taylor, "Planning and Execution of Straight-line Manipulator Trajectories," IBM J. Research and Development 23 1979, p. 424.

M. Hollerbach. "A Recursive Formulation of Lagrangian Manipulator Dynamic" IEEE Trans. Systems. Man. Cybernetics SMC-l0, 11, 1980, p. 730.

J.Y.S. Luh, M.W. Walker, and R.P.c. Paul, "On-line Computational Scheme for Mechanical Manipulators," J. Dynamic S)'stems. Measurement. Control, 102, 1980,

C.S.G. Lee, P.R. Chang, "Efficient Parallel Algorithm for Robot Inverse Dynamics Computation," IEEE Trans. on Systems. Man and Cybernetics, Vol. SMC-16, No.4, July/August 1986,

E.E. Binder, J.H. Herzog, "Distributed Computer Architecture and Fast Parallel Algorithm in Real-Time Robot Control," IEEE Trans. on Systems. Man and Cybernetics, Vol. SMC-16, No. 4, July/August 1986,

M.H. Raibert and J.1. Craig, "Hybrid position/force control of manipulators," J. Dynamic Systems. Measurement. Control, June, 1981, p. 126.

H. Kazerooni, T.B. Sheridan, P.K. Houpt, "Robust Compliant Motion for Manipulators, Part I: The Fundamental Concepts of Compliant Motion," JEEE Journal of Robotics and Automation, Vol. RA-2, No.2, June 1986, p. 83.

H. Kazerooni, P.K. Houpt, T.B. Sheridan, "Robust Compliant Motion for Manipulators, Part II: Design Method," IEEE Journal of Robotics and Automation, Vol. RA-2, No.2, June 1986, p.93.

D.F. Golla, S.C. Garg, and P.c. Hughes, "Linear State-Feedback Control of Manipulators," Mech. Machine Theory, 16, 1981, p. 93.

F. Freund, "Fast Nonlinear Control with Arbitrary Pole- placement for Industrial Robots and Manipulators," Jnt. 1. Robotics Research 1, 1, 1982, p. 65.

W. Hamel and M. Feldman, "The Advancement of Remote Technology: Past Perspectives and Future Plans," Proc.1984 National Topical Meeting on Robotics and Remote Handling in Hostile Environments, ANS, Gatlinburg, TN, April, 1984.

J. Herndon and W. Hamel, "Analysis of the Options=Rationale for Servomanipulators in Future Reprocessing Plants," Proc.1984 National Topical Meeting on Robotics and Remote Handling in Hostile Environments, ANS, Gatlinburg, TN, April, 1984.

W. Hamel et. al., "Advanced Teleoperation in Nuclear Applications," 1984 ASME International Computers in Engineering Conf., Las Vegas, NV. August, 1984. W. Hamel and H. Martin, "Robotics-related Technology in the Nuclear Industry," Proc. SPIE, Vol. 442, August, 1983, pp. 97 -107.

A. K. Bejczy and J. K. Salisbury, "Controlling Remote Manipulators Through Kinesthetic Coupling," Computers in Mechanical Engineering, July, 1983, pp. 48-60.

L. Grisham, "Monitors 1980: Now There are Two," Proc. 28th Conf. Remote Systems Technology, Vol. 2, American Nuclear Society, 1980, p. 83.

J. Albus, C. Mclean, A. Barbera, M. Fitzgerald, " An Architecture for Real-Time Sensory-Interactive Control of Robots in a Manufacturing Environment," 4th IFACIIFIP Symposium on Information Control Problems in Manufacturing Technology, Gaithersburg, Oct., 1982.

J T. Hong and M. Shneier, "Describing a Robot's Workspace Using a Sequence of Views from a Moving Camera," IEEE Trans. PAMI, vol PAMI-7, no. 6,1985, p. 721.

P. Brown, "The Interactive Process Planning System," 1986 Winter ASME Conf., Anaheim, Dec., 1986 (submitted).

W.A. Perkins, "A Model Based Vision System for Industrial Parts," IEEE Trans. on Computers, Vol. C-27, 1978, p. 126.

G.L. Gleason, G.J. Agin, "A Modular Vision System for Sensor-controlled Manipulation and Inspection," Proc. 9th Int. Symposium on Industrial Robots, 1979, p. 57.

M.R. Ward, et.al., "CONSIGHT An Adaptive Robot with Vision, "Robotics Today, 1979, p. 26.

J. E. Kent, M. Nashman, P. Mansbach, L. Palombo, M.A. Shneier, "Six Dimensional Vision System," SPIE, Vol. 336, Robot Vision, 1982, p. 142.

R.C. Bolles, R.A. Cain, "Recognizing and Locating Partially Visible Objects: The Local Feature-Focus Method," Int. Journal of Robotics Research, Vol. 1, 1982, p. 57.

R.C. Bolles, P. Horaud, M.I. Hannah, "3DPO: Three Dimensional Parts Orientation System." Proc. of The 8th. Joint Conf. on Artificial Intelligence, August 1983, p. 1116.

T.F. Knoll and R.C. Jain, "Recognizing Partially Visible Objects Using Feature Indexed Hypotheses," Proc. IEEE Conf. on Robotics and Automation, San Francisco, 1986. p.925.

C. Crowley, "Navigation for an Intelligent Mobile Robot," IEEE Journal of Robotics and Automation, Vol. RA-l, No.1, 1985, p. 31.

R. Lumia, "Representing Solids for a Real-Time Robot Sensory System," Proc. Prolamat 1985, Paris, June 1985.

M.O. Shneier, E.W. Kent, P. Mansbach, "Representing Workspace and Model Knowledge for a Robot with Mobile Sensors," Proc. 7th Int. Conf. Pattern Recognition, 1984. p. 199.

M. A.. Shneier, R. Lumia, E.W. Kent, "Model Based Strategies for High Level Robot Vision," CVGIP, Vol. 33,1986, p. 293.

C. McLean, H. Bloom, T. Hopp, "The Virtual Manufacturing Cell, "IFACIIFIP Cont'. on Information Control Problems in Manufacturing Technology, Gaithersburg, MD., Oct.,1982.

A. Barr, E. Feigenbaum, *The Handbook of Artificial Intelligence*, Kaufman, 1981, ASIN-B01DSRU4UW.

C. McLean, M. Mitchell, E. Barkmeyer, "A Computer Architecture for Small Batch Manufacturing," IEEE Spectrum, May, 1983, p. 59.

M. Mitchell, E. Barkmever, "Data Distribution in the NBS Automated Manufacturing Research Facility", Proc. of the (PAD2 Conf., Denver, April, 1984.

C. McLean, "Process Planning Research at the AMRF," Navy Manufacturing Technology Report, Nov., 1984, p.5.

C. McLean, "An Architecture for Intelligent Manufacturing Control." 1st ASME Computers in Engineering Conf., Boston, Aug., 1985.

If you enjoyed this book, you might also be interested in some of these.

Stakem, Patrick H. *16-bit Microprocessors, History and Architecture*, 2013 PRRB Publishing, ISBN-1520210922.

Stakem, Patrick H. *4- and 8-bit Microprocessors, Architecture and History*, 2013, PRRB Publishing, ISBN-152021572X,

Stakem, Patrick H. *Apollo's Computers*, 2014, PRRB Publishing, ISBN-1520215800.

Stakem, Patrick H. *The Architecture and Applications of the ARM Microprocessors*, 2013, PRRB Publishing, ISBN-1520215843.

Stakem, Patrick H. *Earth Rovers: for Exploration and Environmental Monitoring*, 2014, PRRB Publishing, ISBN-152021586X.

Stakem, Patrick H. *Embedded Computer Systems, Volume 1, Introduction and Architecture*, 2013, PRRB Publishing, ISBN-1520215959.

Stakem, Patrick H. *The History of Spacecraft Computers from the V-2 to the Space Station*, 2013, PRRB Publishing, ISBN-1520216181.

Stakem, Patrick H. *Floating Point Computation*, 2013, PRRB Publishing, ISBN-152021619X.

Stakem, Patrick H. *Architecture of Massively Parallel Microprocessor Systems*, 2011, PRRB Publishing, ISBN-1520250061.

Stakem, Patrick H. *Multicore Computer Architecture,* 2014, PRRB Publishing, ISBN-1520241372.

Stakem, Patrick H. *Personal Robots*, 2014, PRRB Publishing, ISBN-1520216254.

Stakem, Patrick H. *RISC Microprocessors, History and Overview,* 2013, PRRB Publishing, ISBN-1520216289.

Stakem, Patrick H. *Robots and Telerobots in Space Application*s, 2011, PRRB Publishing, ISBN-1520210361.

Stakem, Patrick H. *The Saturn Rocket and the Pegasus Missions, 1965,* 2013, PRRB Publishing, ISBN-1520209916.

Stakem, Patrick H. *Visiting the NASA Centers, and Locations of Historic Rockets & Spacecraft,* 2017, PRRB Publishing, ISBN-1549651205.

Stakem, Patrick H. *Microprocessors in Space*, 2011, PRRB Publishing, ISBN-1520216343.

Stakem, Patrick H. Computer *Virtualization and the Cloud*, 2013, PRRB Publishing, ISBN-152021636X.

Stakem, Patrick H. *What's the Worst That Could Happen? Bad Assumptions, Ignorance, Failures and Screw-ups in Engineering Projects, 2014,* PRRB Publishing, ISBN-1520207166.

Stakem, Patrick H. *Computer Architecture & Programming of the Intel x86 Family, 2013,* PRRB Publishing, ISBN-1520263724.

Stakem, Patrick H. *The Hardware and Software Architecture of the Transputer*, 2011,PRRB Publishing, ISBN-152020681X.

Stakem, Patrick H. *Mainframes, Computing on Big Iron*, 2015, PRRB Publishing, ISBN- 1520216459.

Stakem, Patrick H. *Spacecraft Control Centers*, 2015, PRRB Publishing, ISBN-1520200617.

Stakem, Patrick H. *Embedded in Space,* 2015, PRRB Publishing, ISBN-1520215916.

Stakem, Patrick H. *A Practitioner's Guide to RISC Microprocessor Architecture*, Wiley-Interscience, 1996, ISBN-0471130184.

Stakem, Patrick H. *Cubesat Engineering*, PRRB Publishing, 2017, ISBN-1520754019.

Stakem, Patrick H. *Cubesat Operations*, PRRB Publishing, 2017, ISBN-152076717X.

Stakem, Patrick H. *Interplanetary Cubesats*, PRRB Publishing, 2017, ISBN-1520766173 .

Stakem, Patrick H. Cubesat Constellations, Clusters, and Swarms, Stakem, PRRB Publishing, 2017, ISBN-1520767544.

Stakem, Patrick H. *Graphics Processing Units, an overview*, 2017, PRRB Publishing, ISBN-1520879695.

Stakem, Patrick H. *Intel Embedded and the Arduino-101, 2017,* PRRB Publishing, ISBN-1520879296.

Stakem, Patrick H. *Orbital Debris, the problem and the mitigation*, 2018, PRRB Publishing, ISBN-*1980466483.*

Stakem, Patrick H. *Manufacturing in Space*, 2018, PRRB Publishing, ISBN-1977076041.

Stakem, Patrick H. *NASA's Ships and Planes*, 2018, PRRB Publishing, ISBN-1977076823.

Stakem, Patrick H. *Space Tourism*, 2018, PRRB Publishing, ISBN-1977073506.

Stakem, Patrick H. *STEM – Data Storage and Communications*, 2018, PRRB Publishing, ISBN-1977073115.

Stakem, Patrick H. *In-Space Robotic Repair and Servicing*, 2018, PRRB Publishing, ISBN-1980478236.

Stakem, Patrick H. *Introducing Weather in the pre-K to 12 Curricula, A Resource Guide for Educators*, 2017, PRRB Publishing, ISBN-1980638241.

Stakem, Patrick H. *Introducing Astronomy in the pre-K to 12 Curricula, A Resource Guide for Educators*, 2017, PRRB Publishing, ISBN-198104065X. Also available in a Brazilian Portuguese edition, ISBN-1983106127.

Stakem, Patrick H. *Deep Space Gateways, the Moon and Beyond*, 2017, PRRB Publishing, ISBN-1973465701.

Stakem, Patrick H. *Exploration of the Gas Giants, Space Missions to Jupiter, Saturn, Uranus, and Neptune*, PRRB Publishing, 2018, ISBN-9781717814500.

Stakem, Patrick H. *Crewed Spacecraft*, 2017, PRRB Publishing, ISBN-1549992406.

Stakem, Patrick H. *Rocketplanes to Space*, 2017, PRRB Publishing, ISBN-1549992589.

Stakem, Patrick H. *Crewed Space Stations,* 2017, PRRB Publishing, ISBN-1549992228.

Stakem, Patrick H. *Enviro-bots for STEM: Using Robotics in the pre-K to 12 Curricula, A Resource Guide for Educators,* 2017, PRRB Publishing, ISBN-1549656619.

Stakem, Patrick H. *STEM-Sat, Using Cubesats in the pre-K to 12 Curricula, A Resource Guide for Educators*, 2017, ISBN-1549656376.

Stakem, Patrick H. *Lunar Orbital Platform-Gateway*, 2018, PRRB Publishing, ISBN-1980498628.

Stakem, Patrick H. *Embedded GPU's*, 2018, PRRB Publishing, ISBN-1980476497.

Stakem, Patrick H. *Mobile Cloud Robotics*, 2018, PRRB Publishing, ISBN-1980488088.

Stakem, Patrick H. *Extreme Environment Embedded Systems,* 2017, PRRB Publishing, ISBN-1520215967.

Stakem, Patrick H. *What's the Worst, Volume-2*, 2018, ISBN-1981005579.

Stakem, Patrick H., *Spaceports*, 2018, ISBN-1981022287.

Stakem, Patrick H., *Space Launch Vehicles*, 2018, ISBN-1983071773.

Stakem, Patrick H. *Mars*, 2018, ISBN-1983116902.

Stakem, Patrick H. *X-86, 40th Anniversary ed*, 2018, ISBN-1983189405.

Stakem, Patrick H. *Lunar Orbital Platform-Gateway*, 2018, PRRB Publishing, ISBN-1980498628.

Stakem, Patrick H. *Space Weather*, 2018, ISBN-1723904023.

Stakem, Patrick H. *STEM-Engineering Process*, 2017, ISBN-1983196517.

Stakem, Patrick H. *Space Telescopes,* 2018, PRRB Publishing, ISBN-1728728568.

Stakem, Patrick H. *Exoplanets*, 2018, PRRB Publishing, ISBN-9781731385055.

Stakem, Patrick H. *Planetary Defense*, 2018, PRRB Publishing, ISBN-9781731001207.

Patrick H. Stakem *Exploration of the Asteroid Belt*, 2018, PRRB Publishing, ISBN-1731049846.

Patrick H. Stakem *Terraforming*, 2018, PRRB Publishing, ISBN-1790308100.

Patrick H. Stakem, *Martian Railroad,* 2019, PRRB Publishing, ISBN-1794488243.

Patrick H. Stakem, *Exoplanets,* 2019, PRRB Publishing, ISBN-1731385056.

Patrick H. Stakem, *Exploiting the Moon,* 2019, PRRB Publishing, ISBN-1091057850.

Patrick H. Stakem, *RISC-V, an Open Source Solution for Space Flight Computers,* 2019, PRRB Publishing, ISBN-1796434388.

Patrick H. Stakem, *Arm in Space*, 2019, PRRB Publishing, ISBN-9781099789137.

Patrick H. Stakem, *Extraterrestrial Life*, 2019, PRRB Publishing, ISBN-978-1072072188.

Patrick H. Stakem, *Space Command*, 2019, PRRB Publishing, ISBN-978-1693005398.